AF247629

## GEOP
### *Gas Engineering and Operating Practices*

A Series by the Operating Section
**The American Gas Association**

# Volume II
# TRANSMISSION

## Book T-2
## Compressor Station Operations

The American Gas Association
Arlington, Virginia

# CONTENTS

# FIGURES AND TABLES

## FIGURES

# TABLES

# PREFACE

Arlington, Virginia<br>
February 1985

The Gas Engineering and Operating Practices Series consists of five Volumes comprising twelve separate Books addressing the various topics of gas supply, transmission, distribution, utilization, and related technologies. Contributing authors are veteran practitioners in the gas industry who were selected for their subject knowledge from 22 technical committees in the A.G.A. Operating Section, as well as academics, consultants, interested suppliers, and other specialists.

This *Compressor Station Operations Book* is one of three that will constitute the GEOP Transmission Volume, the other two being *Pipeline/Transmission Planning and Economics* and *Gas Control/Automation and Telecommunications.*

A work of this sort, by its nature, does not afford the detailed treatment possible in more specialized texts. The references provided in the footnotes and the Bibliography expand many-fold the usefulness of the Compressor Book. Readers owe a special debt of gratitude to Florine Hunt and fellow members of the special GEOP Subcommittee of the A.G.A. Library Services Committee, who did a thoroughly professional job of reviewing the manuscript for helpful insertion of footnotes and assured the accuracy of the Bibliography. They also reorganized Appendix A, Codes and Standards, to make it easier to understand and use.

Also to be remembered is the support and wise counsel of the first GEOP Chairman, the late Ty Miller, one of the Series' chief architects and principal champions.

Anthony A. Bobelis      Robert L. Parker  
Chairman, GEOP       Editor  
Task Group

# GEOP TASK GROUP

The following Operating Section members have served on the Gas Engineering and Operating Practices Task Group of the Section Managing Committee through the development of the GEOP Series.

*Anthony A. Bobelis*, Chairman (1982–present), Principal Engineer, Engineering Department, The Brooklyn Union Gas Company

*Ty S. Miller*, Chairman (1979–1982), Manager, Distribution (Deceased), Southern California Gas Company

*C. William Ade* (1980–1983), Vice President, Gas Transmission and Planning, Mississippi River Transmission Company

*Anthony A. Bobelis* (1979–1982), Principal Engineer, Engineering Department, The Brooklyn Union Gas Company

*Alton T. Davis* (1979–1980), Vice President, Gas Division, San Diego Gas and Electric Company

*Robert H. Holder* (1982–present), Vice President and Chief Engineer, Wisconsin Gas Company

*George M. Hugh* (1979–1983), Senior Vice President, Engineering and Operations, TransCanada PipeLines Ltd.

*Vernon E. Percell* (1983–present), Senior Vice President, Division Operations, The Gas Service Company (of Kansas City)

*Marvin D. Ringler* (1979–1981), Engineer, Gas Supply Planning, Consolidated Edison Company of New York, Inc.

*O. L. Slaughter* (1983–present), Senior Vice President, Gas Company of New Mexico

*Robert D. Stegner* (1983–present), Senior Vice President, Operations and Engineering, Indiana Gas Company

*Andrew Tarapchak* (1979–1984), Vice President, Distribution, Washington (D.C.) Gas Light Company

*Milton M. Walther* (1979–1983), General Manager, Gas Department, New Orleans Public Service, Inc.

*Robert L. Parker* (1980–present), Manager, Engineering Series Publications, American Gas Association

# ACKNOWLEDGEMENTS

This book is a particular tribute to the management and editorial skills and efforts of Donald E. Gasser, Task Group Leader of the Operating Section Compressor Committee for the Gas Engineering and Operating Practices Series, and to the following Compressor Committee members and other expert contributors.

**Contributing Authors:** *W. W. Adams, Jr.,* Administrator–Transmission Services, Transcontinental Gas Pipe Line Corporation. *James Adkins,* Manager, Electrical Engineering (Retired), Tennessee Gas Pipeline Company. *Eugene Amburgey,* Staff Engineer, Tennessee Gas Pipeline Company. *James R. Franklin,* Division Engineer–Design Development, Panhandle Eastern Pipe Line Company. *Donald E. Gasser,* Manager–Gas Supply Engineering, The East Ohio Gas Company. *Haim H. Haimov,* Compressor Service Administrator, Southern California Gas Company. *John T. Hood, Jr.,* Assistant General Superintendent–Compressor Stations, United Gas Pipe Line Company. *Barney Inglis,* Civil Engineer (Retired), Tennessee Gas Pipeline Company. *Harvey D. Johnson,* Senior Engineer–Project Design, Panhandle Eastern Pipe Line Company. *James M. Kelly,* Division Engineer–Project Design, Panhandle Eastern Pipe Line Company. *Conrad D. Lawson,* General Superintendent of Transmission, Natural Gas Pipeline Company of America. *Alfred W. Matthes II,* formerly Associate Engineer–Design Division, Panhandle Eastern Pipe Line Company. *Neal A. Matthews,* Senior Compressor Engineer, Southern Natural Gas Company. *Mervyn O. Maxson,* Manager–Area Transmission System, Michigan Wisconsin Gas Company. *Norris E. McDivitt,* Division Superintendent, Texas Gas Transmission Corporation. *Richard R. McGinnis,* Division Superintendent, Southern Natural Gas Company. *Jack Nagel,* Senior Engineer–Design Development, Panhandle Eastern Pipe Line Company. *Don H.*

*Pendleton*, Gas Compression Superintendent, Consumers Power Company. *Heral Singleton*, Senior Compressor Engineer, Tennessee Gas Pipeline Company. *Carl J. Veit*, Superintendent – Compressor Stations (Retired), The East Ohio Gas Company. *Benno W. Walters*, Senior Compressor Engineer, Tennessee Gas Pipeline Company. *K. F. Wrenn, Jr.*, Manager, Compressor Operations and Applications – Engineering, Columbia Gas Transmission Corporation.

Reviewing and integrating the writing of the authors into the finished product was the work of still other contributors in various industry-related fields.

**Cognizant GEOP Task Group Member:** *George M. Hugh*, TransCanada PipeLines Ltd. **Compressor Committee GEOP Task Group:** *Donald E. Gasser*, Task Leader, The East Ohio Gas Company. *W. W. Adams, Jr.*, Transcontinental Gas Pipe Line Corporation. *C. Clarke Kiner*, Columbia Gas Transmission Company. *George F. Mandy*, Southern Natural Gas Company. *Benno W. Walters*, Tennessee Gas Pipeline Company. **Compressor Committee GEOP Review Subcommittee.** *Kenneth W. Corder*, El Paso Natural Gas Company. *Franklin P. Jackson*, Algonquin Gas Transmission Company. *Mervyn O. Maxson*, Michigan Wisconsin Pipe Line Company. *Don H. Sloan* (Retired), Transwestern Pipeline Company. **Editorial and Typing Support:** *Ruth E. Suess*, The East Ohio Gas Company. *John F. Schniegenberg*, The East Ohio Gas Company. **Indexing and Bibliography:** Chairman *Florine E. Hunt*, Public Service Electric and Gas Company, *Mirren Hinchley, Anne C. Roess, Nancy L. Urbankiewicz, Cheryl Watson,* and *Louise Whitaker* of the GEOP Subcommittee of the Library Services Committee, A.G.A. Financial and Administrative Section. *Susan T. Bau*, American Gas Association. **Metric Review:** *D. G. Donaghey*, TransCanada PipeLines Ltd., and members of the Metrication Task Committee of the A.G.A. Operating Section. **Accident Prevention Review:** *W. H. Robins*, Panhandle Eastern Pipe Line Company, and members of the Accident Prevention Committee of the A.G.A. Operating Section. **Cover Design:** *Jeanine Banks*, American Gas Association.

# OPERATING SECTION
# AMERICAN GAS ASSOCIATION

## OFFICERS

Frederick W. Sullivan
Chairman
The Brooklyn Union Gas
  Company
Brooklyn, New York

Andrew C. Olson
Second Vice Chairman
Minnegasco, Inc.
Minneapolis, Minnesota

George F. Mandy
Fourth Vice Chairman
Southern Natural Gas Company
Birmingham, Alabama

Louis C. Baldacci, Jr.
First Vice Chairman
Peoples Gas Light & Coke
  Company
Chicago, Illinois

William L. Clayton
Third Vice Chairman
ENTEX, Inc.
Houston, Texas

Roy A. Siskin
Section Staff Executive
American Gas Association
Arlington, Virginia

## A.G.A. STAFF

Louis A. Sarkes
Staff Vice President,
Engineering

John P. Erickson
Manager, Engineering
  Services Programs

Larry T. Ingels
Manager, Engineering
  Services Programs

Jimmie A. Hicks
Manager, Engineering
  Services, Programs

Robert L. Parker
Manager, Engineering
  Series Publications

Philip S. Runge
Manager, Safety and
Insurance Services

# Chapter 1

# INTRODUCTION

A compressor station is one of the most critical components in a natural gas pipeline system. Compressor stations supply the energy to pump gas from production fields, to overcome frictional losses in transmission pipelines, and to pump gas into and from storage reservoirs. Careful design and operation of these facilities are vital to maintaining the integrity of the entire pipeline system.

## SCOPE AND OUTLINE

The intent of this Book is to provide the entry level engineer, the resident compressor designer, and other gas industry personnel, particularly those engaged in gas supply projects, with the fundamentals of natural gas compression from the planning through the operating phases. The following topics are covered:

- Principal design considerations common to all types of stations concerning site selection, buildings, foundations, prime movers, gas compressors, piping, auxiliary equipment, and instrumentation
- Basic analytical, thermodynamic relationships applied to the gas compression process
- Personnel requirements, training programs, equipment performance testing, and other aspects associated with the operation of compressor stations
- Routine and preventive maintenance programs including such topics as electronic analysis programs, lube oil analysis, and performance-monitoring of machinery
- Descriptions of the various types of safety control equipment and

programs for protecting plant equipment and maintaining a safe working environment

- Economic decisions relating to the selection of compressor station equipment, purchased versus leased equipment, and capital investment versus operating costs
- Pertinent codes and regulations that regulate the design, construction, and operation of compressor stations.

Inherently, the design, engineering, operation, and maintenance of non-conventional compressor facilities, such as those used in offshore platforms, are similar to those for land-based stations. However, because of the unique considerations relating to physical characteristics and special code requirements, this text will not cover the specifics of those types of installations. Ordinarily the design and engineering of such facilities are done by engineering firms that specialize in the specific field of application. For instance, an offshore compressor system commonly is an integral part of the entire offshore platform design, and usually involves an entirely different set of design criteria than onshore stations require. Fuel economy often is sacrificed for size and weight advantages; convenience of maintenance and repair is sacrificed for conservation of space; and code requirement as relating to safety often is more stringent for offshore facilities.

Although every effort was made to include the most commonly accepted standards and practices for design, construction, operation, and maintenance of conventional natural gas compressor stations, it is difficult to cover all contemporary methods in this type of publication. Nevertheless, the fundamental principles included in this section have served the gas industry for decades and will continue to provide the basis for the design and operation of compressor facilities in the future.

## TYPES OF STATIONS

Depending on its purpose, a compressor station generally falls into one of three distinct classifications – production, storage, or transmission. Despite similarities in the above three classifications, each application has unique characteristics.

### PRODUCTION STATIONS

Production stations, also referred to as gas-gathering stations, are used in the natural gas production field to pump gas from the wellhead to the pipeline system. Depending on the nature of the production field and the interconnecting pipeline network, economics usually determine the most practical location for a compressor station. One larger station usually has lower annual costs than several smaller stations. However, since it is more economical to transport gas at elevated pressures, long low-pressure suction lines from the wellhead to a single, centrally located compressor station often result in a more costly system than one in which the compressors are located near the wellheads. The logical objective is to achieve an economical balance between consolidating wells in a production field gas gathering system and minimizing the distance between the wellhead and the compressor.

Production stations usually are designed for unattended operation with a life expectancy comparable to that of the production field. Unitized, skid-mounted, medium- to high-speed, packaged compressor units in the 50- to 1 000-hp (35- to 745-kW) range often are utilized for this type of service. They are advantageous because of their availability, ease of installation and relocation, and relatively low installation costs. Frequently production stations are designed and installed within a short time schedule; therefore, decisions concerning design must be predicated on availability of major components.

Figure 1 illustrates a typical, small, low-cost production station featuring two simple, skid-mounted, belt-driven compressor units. Although an installation of this type often satisfies the basic requirements, environmental standards or operational conditions may dictate that extra consideration be given to noise abatement, safety, reliability, air quality, and other sometimes costly factors. Figure 2 shows a more complex station in which equipment is located in buildings and operated by remote control from a gas dispatch center.

Operating pressures in production stations vary widely. The suction pressure may be required to go subatmospheric, while the discharge pressure could be as high as 1 000 lbf/in$^2$ (6 900 kPa) – in which case the gas may be discharged into a high pressure transmission pipeline system.

## STORAGE STATIONS

Storage compressors are used primarily to pump gas from a transmission pipeline into an underground gas storage field during the injection period. In addition to injecting gas during the storage season, some are used to pump gas from the storage field to the pipeline during the withdrawal period. Most storage field compressors are designed to operate over a broad range of volumes and pressures. Because of widely varying conditions, storage stations are generally the most com-

**Figure 1.** Small gas gathering compressor station with two belt-driven compressors

**Figure 2.** Modern Semiautomatic gas gathering compressor station with three skid-mounted reciprocating compressors

plex, both technologically and operationally, normally requiring multiple staged compression, complicated piping arrangements, and complex control equipment.

Storage compressor stations usually are more permanent, more efficient, and definitely more costly than production stations of equivalent capacity. A typical storage station would contain several reciprocating compressor units ranging in size from 1 000 to 6 000 hp (745 to 4 475 kW) each. Occasionally turbine-driven centrifugal compressors are used in storage stations; however, centrifugal compressors are not always as suitable for this type of service as reciprocating compressors because of the broad range of operating conditions imposed on the equipment.

## TRANSMISSION STATIONS

The transmission station is the most common type. Transmission stations, often referred to as pipeline stations, generally contain several large compressor units at one location, each having a power rating of 1 000 to 20 000 hp (745 to 14 900 kW). Located at various points along a transmission pipeline, their purpose is to move gas from one station to the next. While operating pressures seldom exceed 1 000 lbf/in$^2$ (6 900 kPa), modern technology and economics have made it possible to use higher pressures.

A transmission station may utilize reciprocating engines, turbines, or electric motors to drive either reciprocating or centrifugal gas compressors. Designed for long life (50 years or more) and intended to operate year-round, these facilities usually are located on tracts of 20 acres (8 hectare) or larger, with permanent structures to house the compressor units and auxiliary facilities. In general transmission compressor stations are similar in appearance to that shown in Figure 3.

*Courtesy of Tennessee Gas Pipeline Company*

*Figure 3.* **Typical transmission compressor station**

In addition to containing the compressor facilities, the transmission station often is used as a divisional reporting location and supply warehouse for operating personnel assigned to a segment of the transmission pipeline and related facilities.

# Chapter 2

# ENGINEERING AND DESIGN CONSIDERATIONS

Included in this chapter are some of the major factors that should be considered in the engineering and design of conventional land-based natural gas compressor stations. The principal topics discussed are the general physical arrangement of compressor station facilities, an explanation of the various types of prime movers and compressors, a brief theory of compressor design, and a review of the associated equipment and facilities normally provided.

## SITING AND LOCATION

The function of a natural gas pipeline is to transport gas from one geographical location to another. Generally a pipeline will take the most direct route between these two points. The pipeline route determines the location of compressor stations.

### SITE SELECTION

Normally the preliminary compressor station site selection is based on a design analysis of the total pipeline system as an operating unit. Some design factors which influence compressor station locations along the pipeline system are pipe diameters, length of loop lines, operating pressures, and proposed compressor station power.

For instance, the required system capacity could be obtained by installing a specific size of pipe with compressor stations of a certain rating spaced at fixed intervals. This same capacity could be attained with compressor stations spaced farther apart if either the size of the pipe or the station rating was increased. Conversely, stations would have to be spaced closer together if the pipe size was decreased or the station power was reduced.

The designer must examine the different system-design alternatives and choose one that is economically feasible. Normally this will determine compressor station spacing and thus the general location of each station.

Once the initial location on the pipeline is determined, other factors, such as specific geographic location, topography, and environmental considerations, must be evaluated. It is possible that some of these factors will prohibit stations from being built at locations indicated by the initial system evaluation. In such cases it is necessary to modify the system design to accommodate these constraints.

## Geographic Location

Once the general location of the station on the pipeline is determined, a study is initiated to determine the best geographical location. Ideally the station would be located near a railroad siding since large compressor engines usually are shipped by rail. This is not always possible, and it is then necessary to move the engines by truck from the railhead to the station site.

The station should be located near an all-weather road. Even if compressor engines are shipped by rail directly to the station, there probably will be heavy material such as coolers, piping, valves, scrubbers, and miscellaneous vessels that will be shipped by truck. Heavy equipment, such as bulldozers and cranes, will be trucked to the site during construction. The road should have a solid bed, strong bridges, adequate width, and sufficient overhead clearance.

It is preferable to have the station located in a rural rather than an urban area. One advantage is lower site cost, and another is that a rural location will minimize complaints from residents about the noise and vibration normally associated with large reciprocating compressor stations. However, if the location is extremely isolated, additional expenses can be incurred because it may be necessary to construct residential buildings for station personnel. It also may be difficult to hire station personnel if good schools, shopping centers, and recreational facilities are not available.

It is helpful, but not absolutely essential, to have such public utilities as electricity, water, garbage service, and sewerage service available. If these utilities are not available, electricity can be produced using generators driven by natural gas engines. Water wells can be drilled. Some garbage can be buried or burned, but the waste from some products used at compressor stations may be classified as "hazardous waste" by government regulation and require special handling and disposal in approved landfills. Sewage treatment facilities can also be installed. These are not major problems but are factors related to the geographical location, and must be considered when designing a compressor station.

## Topography

Topographic features and the underlying soil characteristics must be considered in designing a compressor station. Ideally the site should be higher than the surrounding area to promote good drainage, and reasonably level to minimize the amount of earthwork. The soil should have good bearing characteristics to minimize the need for special foundations.

Often the site will not have these ideal features, and the designer must make provisions to overcome adverse topographic features. If the site is located in a low area, for example, it will be necessary to provide fill material to elevate and protect critical equipment from flood damage. Special designs incorporating ditches, culverts, pipes, catch basins, and sump pumps may be necessary to ensure proper drainage during heavy rainfall.

If the site is not level, it may be necessary to move earth from high areas to fill the low areas. Such cut-and-fill work can be very expensive if there are extreme elevation differences or if a rock formation is present. Another problem is that fill areas normally have poor foundation characteristics and may be subject to consolidation over the years, resulting in settlement which causes equipment alignment problems.

Foundation design for a compressor station depends on surface and subsurface soil bearing characteristics. Foundations are critical because they must support heavy vibrating loads and must maintain equipment alignment. An analysis of core samples obtained by soil boring is necessary to determine the proper foundation for anticipated loads, both static and dynamic, under both dry and saturated soil conditions. Simple, concrete spread foundations may be adequate in some locations, while other areas may require extensive piling and soil stabilization.

## Environmental Considerations

Public interest and government regulations have made environmental considerations an important factor in compressor station design. Many of the regulations enforced by the federal Environmental Protection Agency (EPA) and similar state environmental agencies can be satisfied through the utilization of sound engineering design.

For instance, the EPA Spill Prevention Control and Countermeasure

Plan requires that measures be taken to ensure that petroleum products will not be spilled into navigable waters or drainage ditches, etc. Current requirements can be met by constructing impervious dikes around storage facilities, and preparing and documenting a written emergency plan of action to be followed in case of a spill. Dikes should be sized to contain the entire contents of the largest single tank plus sufficient freeboard to allow for precipitation.

Examples of other EPA regulations that normally can be handled by sound engineering designs are those that concern the quality of water discharged from any point in the station and regulations pertaining to the disposal of hazardous wastes.

Federal standards for ambient air quality are usually the most critical environmental regulations related to compressor station construction. The Clean Air Act of 1970 as amended in 1977 directs the EPA to establish and enforce standards for ambient air quality to protect public health. Standards have been established for aerial pollutants, including nitrogen oxides ($NO_X$), carbon monoxide (CO), sulfur oxides ($SO_X$), and hydrocarbons (HC). Standards that apply to the emission of nitrogen dioxide ($NO_2$) from exhaust stacks must be considered in compressor station construction. For large compressor stations a Prevention of Significant Deterioration (PSD) Permit must be obtained prior to construction in a clean area. An offset permit is required for noncompliance areas.

There are numerous other environmental regulations that affect the compressor station location and design. Because of the volume and complexity of these regulations, most companies employ environmental specialists to ensure compliance with these regulations and to secure the appropriate permits.


## LAYOUT AND ARRANGEMENT

In 1970 the Department of Transportation (DOT) Material Transportation Bureau (MTB) Office of Pipeline Safety Regulation (OPSR) issued regulations pertaining to compressor design under Subpart D of Part 192, Title 49 of the Code of Federal Regulations, "Transportation of Natural and Other Gas by Pipeline: Minimum Federal Safety Standards" (See Appendix A). Most of the regulations concerning the layout and arrangement of compressor stations are written in terms of performance standards rather than detailed design criteria. This allows the individual designer considerable latitude in the layout and arrangement of stations.

## Size of Station

One way to compare the size of compressor stations is to compare their power ratings. This is an accepted industry practice since compression power is related to the volume of gas that can be compressed. Small stations may have only several hundred horsepower while large stations often have horsepower in multiples of ten thousand. However, the amount of horsepower at a station does not reflect necessarily the physical size of the station. For example, one station may have a single compact 2 500-hp (1 875-kW) turbine unit while another may have five 500-hp (375-kW) reciprocating units, each of which is physically as large as the turbine unit.

Some provisions of the DOT regulations have an effect on the physical size of compressor stations. For example, it is required that each main compressor building be located on property under control of the operator. Such a building must be far enough away from adjacent property to minimize the chances of fire spreading to it from structures located on the adjacent property. Also, there must be enough open space around compressor buildings to allow movement of fire-fighting equipment.

## Other Facilities

Other required facilities can have a significant impact on the physical size of a station.

The following facilities are common to most large compressor stations:
- Air compressors and air storage tanks for engine starting and instrumentation
- Electric power generators
- Fuel gas regulators
- Meters and heaters
- Engine water and oil heat exchangers
- Oil storage tanks
- Gas separators or scrubbers.

The following facilities often are incorporated in compressor station design, depending on individual requirements:
- Main gas flow measurement facilities
- Gas analyzing equipment
- Discharge gas coolers
- Waste material incinerators
- Sump pumps

- Water wells and water storage and treatment facilities
- Oil skimmers
- Electrical transformers and motor control centers
- Special storage facilities for paints and other combustibles
- Office and employee facilities
- Warehouse area
- Workshop area.

## Provisions for Expansion

It is expected that new sources of gas will become available as gas exploration and production technology advances. It also is expected that the demand for natural gas will increase as the population grows. Therefore, provisions should be made for future expansion. Some of these provisions may involve additional costs during initial construction. Examples of these provisions include the installation of:

- Oversized gas piping
- Spare engine capacity
- Foundations and the extension of buildings to receive additional units
- More or larger air compressors
- Scrubbers or separators with excess capacity
- Oil storage facilities with additional capacity.

Other provisions can be made for future expansion that will incur relatively minor additional costs. For instance, usually it is more economical to purchase land for future expansion at the time of the initial purchase. Areas should be allocated for future expansion in a logical manner—for example, the area to install a future oil storage vessel should be located next to the existing oil storage facilities. Compressor buildings, warehouses, and other structures should be designed and located so that they can be expanded with minimum interference to the existing facilities. Suction headers, discharge headers, and other piping that is expected to be extended should terminate with blind flanges to make possible future connections with minimum expenditures. Each station is unique and should be considered individually since future expenditures can be reduced by making provisions during initial design.


# BUILDINGS

The number and size of buildings at compressor stations will vary according to the total prime mover power rating, engine types, compressor types, and the degree of automation. The following discussion will be directed toward large stations.

## BUILDING UTILIZATION REQUIREMENTS

Normally several buildings are used to enclose the machinery and equipment needed to operate and maintain a compressor station. The buildings and the enclosed equipment should be efficiently organized.

Building arrangement varies with the individual designer. In a typical installation the largest structure is the compressor building which contains engines, compressors, suction bottles, discharge bottles, hydraulic pumps, oil filters, water surge tanks, oil surge tanks, fuel gas volume bottles, turbochargers, and other miscellaneous equipment. Instrumentation and controls may be installed on the engines, in separate control panels inside the compressor building, in panels located in a remote control room in a separate building, or in various combinations of these locations.

Air compressors and electric power generators are usually located in a separate structure. Air piping and electrical conduit are routed to the compressor building and other locations as required.

The maintenance building houses a work area and machine shop equipment. This equipment may include a lathe, drill press, hydraulic ram, bench vise, bench grinder, valve stem grinder, power hacksaw, arc welding machine, oxyacetylene equipment, and hand tools.

Flammable materials, such as paint and cleaning supplies, are stored in a noncombustible structure located a safe distance from the other facilities.

Office facilities are provided for the supervisors and clerks. These facilities also may incorporate a personnel locker area, a personnel lunchroom, and an area for meetings. A warehouse should be provided to store spare parts and miscellaneous supplies. Warehouse facilities may be located in the office building or may be combined with the maintenance, auxiliary, or control building.

**NOTE:** *Consider designating a portion of the storage area for controlled items and limiting access to authorized personnel only.*

## BUILDING DESIGN REQUIREMENTS

The design of a main compressor building varies with engine type and size, auxiliary equipment, geographic location, and company philosophy.

**Building Sizing**

The four main types of engine and compressor combinations are reciprocating engines with centrifugal compressors, reciprocating engines with reciprocating compressors, gas turbine engines with centrifugal compressors, and steam turbines with centrifugal compressors. Generally reciprocating engines with reciprocating compressors require the most floor space. Building heights must be sufficient to allow clearance for overhead cranes.

Main compressor building size also depends on operating floor design. Some buildings have either elevated operating floors or full basements. This reduces the amount of required floor space since fuel gas regulators, oil filters, and other engine-related equipment can be located below the operating floor. It should be noted that DOT regulations require that each main compressor building operating floor have two separate exits.

The location of instrumentation and controls will influence the building size. Individual control panels may be installed near each engine, in a control room within the compressor building, or in a separate building.

When sizing the compressor building, adequate space should be provided to perform maintenance work. Ample space must be allowed to remove and repair power cylinders, compressor cylinders, turbine assemblies, centrifugal compressor wheels, and other equipment components. Space should be allocated for storage of maintenance equipment such as tools, jacks, and engine analyzers.


**Climate**

Climate is another factor that can influence building design. In hot climates special provisions should be made to provide proper ventilation. This is especially important because both engines and compressors generate heat. In cold climates it is necessary to provide some form of heat for operating and maintenance personnel. In extremely cold climates the effects of the frost line should be considered in designing the building foundation and services. Figure 4 is a diagram that shows the maximum frost penetration expected in the United States. Similar data is available for other countries also.

In areas of heavy rainfall it may be necessary to install special drainage systems. The operating floor may be raised to a higher elevation to minimize flood damage. This is accomplished by using earth fill to build up the site, or by extending the foundation walls to a higher elevation.

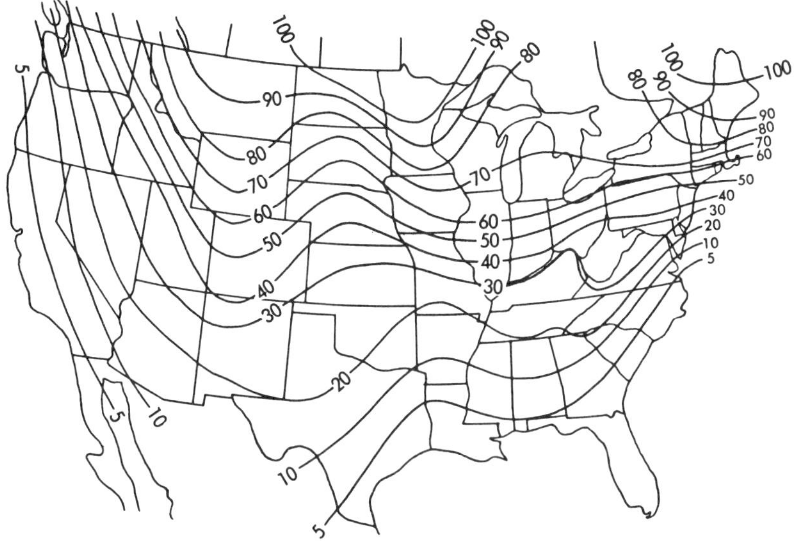

a. Frost Penetration in Inches

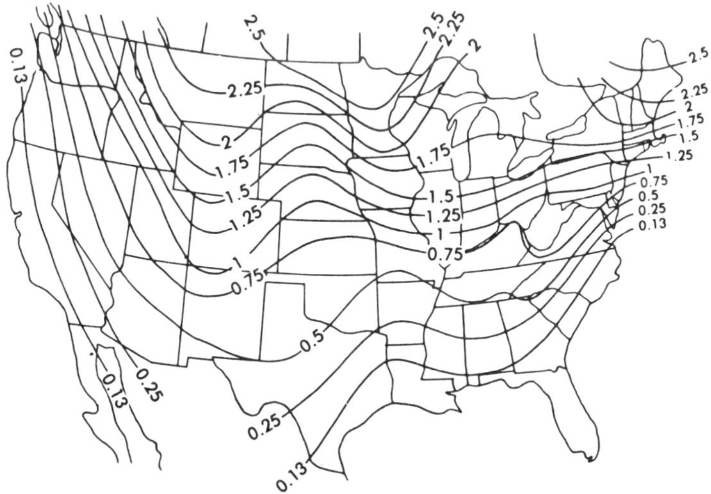

b. Frost Penetration in Metres

***Figure 4.* Maximum frost penetration in the United States**

*a. From* SOIL MECHANICS AND FOUNDATIONS, *Third Edition, by George B. Sowers and George F. Sowers. New York: Macmillan Publishing Company, 1970*
*b. From* INTRODUCTORY SOIL MECHANICS AND FOUNDATIONS: GEOTECHNICAL ENGINEERING, *Fourth Edition, by George F. Sowers. New York: Macmillan Publishing Company, 1979.*

Basement walls should be waterproofed carefully. Basement floors and pipe trenches should be sloped to drain into sumps to minimize water damage caused by flooding or seepage. Pumps may be required to remove water accumulations. Heavy rains may saturate the ground and reduce the soil bearing capacity. This should be considered in the building foundation design.

Special design factors should be considered in areas subjected to high winds or heavy snow loads. The foundation can be strengthened with piling; main structural supports can be crossbraced; the roof design can be strengthened; and provisions can be made to cover glass windows with a protective barrier.


## Environment

Certain environments have an effect on compressor building design. Corrosion is a serious problem in salt water environments. Noncorrosive materials should be used where practical. Steps, handrails, and other components constructed of galvanized steel or aluminum can reduce maintenance. Structural steel should be cathodically protected. Access should be provided to inspect all gas piping routed through building walls.

Extensive piling and special foundation designs are necessary in marsh environments and other areas with poor soil-bearing characteristics.

In areas that have severe dust or blowing sand, it may be necessary to install facilities to clean the air and maintain interior building pressure slightly above the atmospheric pressure.

The environment can affect compressor building design, and conversely, compressor building design can affect the environment. Compressor buildings should be designed, as economically feasible, to minimize their impact on the environment.

The appearance of the compressor building may be a factor in residential areas. Some companies have used brick buildings to house compressor units. These buildings have a structural steel frame to support crane rails. Another environmental concern in residential areas is the noise emitted from engines and compressors. Special air inlet silencers, exhaust mufflers, and gas blowdown silencers can be installed. The most common way to reduce the emission of noise from an engine room to the outside environment is to line the building interior with sound-absorbing material. This also reduces noise inside the building and provides a better working environment. Other methods to reduce noise both inside and outside the engine room consist of the use of sound-absorbing, double cones suspended from the ceiling, and the use of sound barriers or enclosures around each individual engine and compressor unit. These sound barriers should have provisions for forced air ventilation and barrier removal during engine or compressor repairs.

## Construction Materials

The DOT regulations require that each building at a compressor station be constructed of noncombustible materials if it contains pressurized gas piping larger than 2 inches (50 millimetres) nominal diameter, or any gas-handling equipment other than that used for domestic purposes.

It is common practice to construct the building foundation out of reinforced concrete. Wood or concrete pilings are used to support building and equipment foundations when the soil bearing capacity is not sufficient.

Most compressor station buildings are rigid steel, frame structures with transite or corrugated metal siding and roofing. It is relatively easy to adapt this type of building to the installation of traveling cranes and hoists that are used for maintenance work. These buildings can be extended or relocated more readily than other types and generally are accepted as the most economical for large compressor stations. A cross-sectional view of a typical compressor building is shown in Figure 5.

Glass wool material covered by perforated sheet metal or aluminum often is used to insulate engine rooms. This practice reduces both inside and outside noise levels.

Checkered plate or grating is used to cover pipe trenches and other openings in the operating floor.

**CAUTION:** *High humidity makes certain floor surfaces slippery.*

The DOT regulations specify that all electrical facilities in compressor stations must conform to applicable sections of the National Electrical Code (NEC), American National Standards Institute (ANSI) Standard ANSI/NFPA 70 (formerly, ANSI C1). (See Appendix A.)

## Safety Considerations

When a building houses engine-driven units compressing a flammable gas, the National Fire Code (NFPA 37, Section 2–5.1.1) requires provisions to be made for venting an explosion with minimum structural damage. NFPA 68, *Guide for Explosion Venting*, should be consulted.

The DOT regulations also specify that each compressor station building which contains gas-handling equipment must be ventilated to

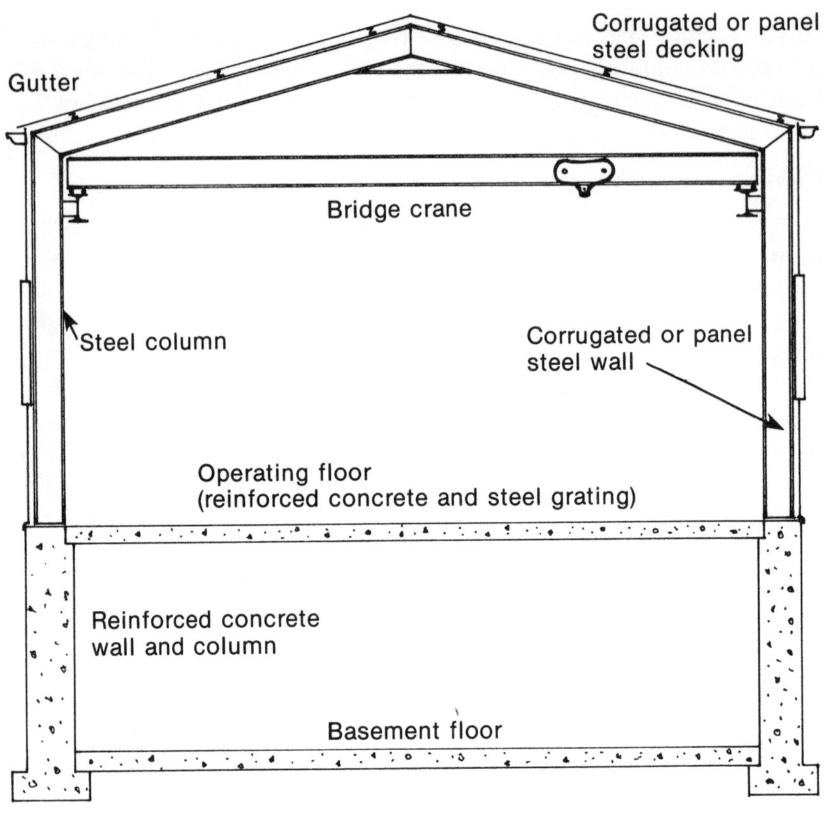


*Figure 5.* **Cross-section of a typical compressor building**

ensure that employees are not endangered by the accumulation of gas
in the enclosed spaces. Same regulations specify that all electrical
facilities in compressor stations must conform to applicable sections
of the National Electrical Code (NEC), American National Standard
ANSI/NFPA 70 (formerly ANSI C1, See Appendix A).

Those areas, in which the electrical apparatus and wiring are judged
to be subject to special requirements outlined in Chapter 5 of the NEC,
are known as hazardous locations. Generally, in a compressor station
only those areas containing gas-handling equipment with significant
leakage potential have to be examined; low pressure domestic gas, or
gas piping without valves, checks, meters and similar devices, would
not ordinarily be deemed to introduce hazardous conditions irrespec-
tive of the operating pressure. In determining the degree of hazard in-

volved, each room or area is evaluated individually. Buildings that house natural gas engines and compressors are subject to classification as hazardous locations by the authorities having jurisdiction; consequently, open flames are not permitted and all electrical facilities must meet the requirements for such locations.

The explosion characteristics of air mixtures of gases, vapors or dusts vary with the specific materials involved. Criteria for determining the suitability of equipment for installation in a given flammable atmosphere include maximum explosion pressure, maximum safe clearance between parts of a clamped joint in an enclosure, and minimum ignition temperature of the atmospheric mixture. Therefore, only electrical equipment marked to show the Class and Group for which it is approved should be selected for installation within the classified area. (In case of heat-producing equipment, the marking shall include the operating temperature or the temperature range identification numbers.) Hazardous areas, the atmospheres of which may contain flammable mixtures of air and hydrocarbon vapors or gases equivalent in hazard to natural gas, are assigned Class I, Group D designation.

More discussion relating to the electrical requirements in the design of compressor station facilities can be found under Electric Power and Lighting, Hazardous Locations.


## Auxiliary Equipment

A great deal of auxiliary equipment is required to operate and maintain large natural gas engines and compressors. Some of this equipment, such as air compressors, generators, and machine shop equipment, usually is located in a building other than the compressor building.

Often other auxiliary equipment is kept in the compressor building so that it will be accessible. These consist of engine analyzers, ladders, various hand tools, and portable fans.

Some equipment is installed permanently in the compressor building – such as gas and fire detection equipment, overhead traveling cranes, hoists, fire extinguishers, ventilation fans, sump pumps, and emergency lighting facilities.


## HEATING AND VENTILATION

In cold climates, buildings in which personnel work for extended periods of time should be heated to enhance productivity and personal comfort.

Compressor buildings commonly are heated with steam or hot water. The boiler is fired by natural gas and located a safe distance from the compressor building. Steam or hot water is piped to the building where it is routed through heat exchangers. Fans are used to circulate air across the heat exchangers and throughout the building.

It is essential that compressor buildings be properly ventilated. This helps minimize the potential for dangerous gas accumulations and also helps remove heat generated by engines and compressors. Compressor buildings equipped with gas detectors may be tied into the wall vent louver controls to help ventilate buildings under certain conditions.

Manufacturers of ventilation equipment have suggested an air change every 1 to 2 minutes for compressor buildings. Ventilation may be supplied by roof ridge vents, unpowered turbine roof ventilators, permanently mounted wall fans, portable cooling fans or fixed wall louvers.

Often combinations of these methods are used. For example, inlet wall fans may be located low on the building wall to draw in cool air. Roof vents and exhaust fans may be located high in the building to discharge hot air.

The maintenance building, warehouse, and office building normally are located in non-hazardous locations. These buildings may be heated by steam, hot water, gas, electricity, kerosene, or infrared heaters. It should be noted that natural gas used for domestic purposes should be odorized.

Buildings located in non-hazardous locations may be ventilated in the same manner as buildings located in hazardous locations, but at a lower rate of air changes. Air change rates of one every 3 to 6 minutes for a machine shop, 2 to 8 minutes for an office, and 10 to 15 minutes for a warehouse are suggested.


# YARD STRUCTURES

The proper design of yard structures is necessary to promote efficient compressor station operation and maintenance.


## ROADS AND SIDEWALKS

Compressor station roads are designed with a base course which includes various combinations of compacted gravel, crushed stone, sand, shell, clay, or stabilized soil. Roads may be surfaced with gravel, crushed stone, shell, asphalt, or concrete depending upon the maintenance and service requirements. Roads must be designed to support vehicles delivering heavy loads of oil, engine components, and other equipment to the station.

The surfaced width should be 10 to 14 feet (3.0 to 4.3 metres) for single-lane roads, and 20 to 24 feet (6.1 to 7.3 metres) for two-lane roads. Concrete and asphalt roads should have a crown at the centerline and should be sloped about 2 percent to ensure proper drainage. Ideally these roads should have a 8- to 10-foot (2.4- to 3.0-metre) wide shoulder on each side.

Parking areas also must be provided for employees and visitors. These areas are constructed of the same materials as the roads.

Roads and parking areas should be designed to facilitate future expansion when there is a possibility that new engines and large cranes may be trucked into the station. Therefore, side and overhead road clearances should not be restricted.

Sidewalks should be wide enough to accommodate handcarts that may be used to move equipment between locations. Two feet ( 0.6 metre) is the design width for one person, and 4 feet (1.2 metres) is the design width for two people walking abreast. Normally sidewalks are constructed of concrete with a wire mesh reinforcement. The thickness of the concrete may vary from 3 to 6 inches (75 to 150 millimetres). Wide sidewalks should be sloped slightly to promote good drainage.


## FENCING

Fences around compressor stations establish the operating boundary, discourage the entry of unauthorized personnel, and promote general security.

Normally chain link fencing is used to establish a perimeter around the entire station facilities. The office and warehouse facilities may be located either inside or outside the perimeter fence. Some designers feel that plant security is enhanced by using a fence to separate these facilities from the station.

DOT regulations require that each fence around a compressor station must have at least two gates, or other exit facilities, conveniently located to provide an opportunity for escape to a place of safety. Gates that are located within 200 feet (61 metres) of any station building must open outward, and when the station is occupied, these gates must be accessible from the inside without a key.


## CULVERTS AND DRAINAGE

Proper drainage is necessary to ensure accessibility of station facilities and to prevent equipment damage because of flooding. The station yard must be properly sloped to direct runoff to drainage ditches and catch basins. Catch basins, underground drainage pipe, sump pumps, and drainage ditches should be sized to accommodate the anticipated peak water runoff.

The anticipated peak water runoff is based on several design parameters. The main parameter is the one-hour design rainfall, which is the amount of rain that is expected to occur once in a specified time frame — usually from 10 to 25 years. Another design variable is the coefficient of runoff which may vary from almost 100 percent for roofs to 10 percent for loose, sandy soil with dense vegetation. The watershed area and the time required for runoff to flow from the most remote point to the discharge point also should be considered.

Most large compressor stations have storm drains with catch basins and underground piping in the valve yard. Roof drains often are tied into this system. Water may flow by gravity to a discharge ditch. However, in flat areas it may be necessary to use sump pumps to lift this water into the discharge ditch. Clear areas may be sloped to drain into open ditches. Metal or concrete culverts are installed under roads and sidewalks. It is essential that road culverts be designed to withstand future vehicular loads.

One special problem is the drainage from petroleum product storage areas. To comply with EPA regulations, these areas usually are diked if the storage facilities have an aggregate capacity of more than 1 320 gallons (4 997 litres), or a single container with a capacity in excess of 660 gallons (2 498 litres). The dike must be sized to contain the entire contents of the largest single tank plus sufficient freeboard to allow for precipitation. Rainwater can be removed from this diked area through the use of gravity drains with isolation valves, or by manually operated sump pumps. Procedures must be implemented to ensure that drain valves are closed and sump pumps are turned off except when unpolluted rainwater is being drained.

# PRIME MOVERS

Various types of prime movers have been used to drive natural gas compressors — the most conventional utilizing natural gas as their source of fuel. Currently the most common types in use include internal combustion, reciprocating engines, gas turbines, and in some cases electric motors. Others, which are not quite as popular but are mentioned in this chapter, are rotary engines and steam turbines. A comparison of thermal efficiencies for several types of prime movers is illustrated in the diagram in Figure 6.

### RECIPROCATING ENGINES

Reciprocating engines used by the gas industry to drive gas compressors are usually spark-ignited, 2-stroke and 4-stroke engines using

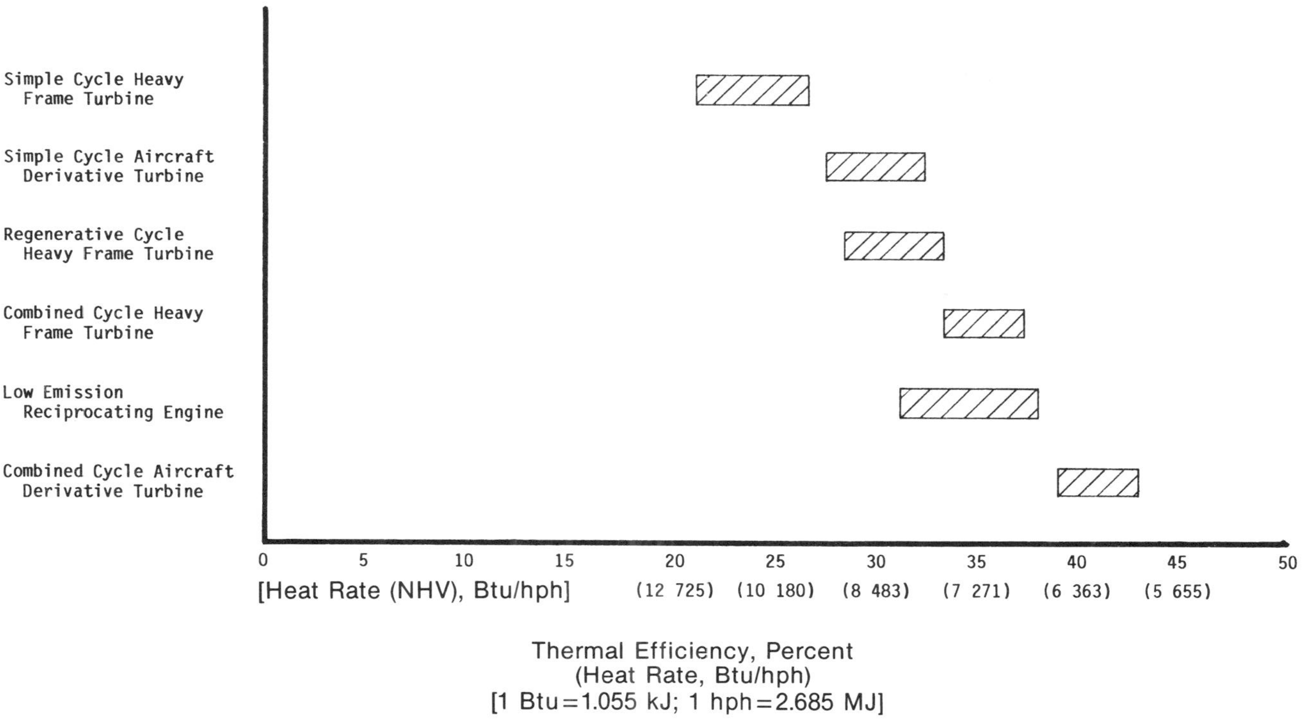


***Figure 6.* Fuel Efficiency for various prime mover arrangements**

*From Mark Axford and Dan Stinger, "Combined Cycle Operation for Pipeline Compressor Stations,"* Turbomachinery International, *July/August 1980*

natural gas for fuel. Diesel-fueled engines would be used only under unusual circumstances because of the unavailability of high-quality natural gas for fuel. The most common type of engine/compressor used is termed integral because the gas compressors are an integral part of the engine, such that the reciprocating compressor connecting rods are attached to the engine crankshaft. Some installations use external compressor drives such as separate crankcase frames with reciprocating compressors, and speed increaser gear assemblies with centrifugal compressors. Both 2-stroke and 4-stroke engines may be equipped with a turbocharger that increases the unit power output and thermal efficiency. Reciprocating engines with a thermal efficiency as high as 40 percent are available. Cross-sectional views of both 2-stroke and 4-stroke gas engines are shown in Figures 7 and 8.

Since engines are rated in different ways, caution must be used when selecting equipment to ensure that the engine is properly rated to meet the power requirements. Slow speed engines usually are rated for continuous duty. High speed engines may be rated for peak output or intermittent duty. The required power to drive all accessories, such as pumps, generators, and fan drives, must be subtracted from the gross power to determine the actual power available to drive the compressors.

### Types of Reciprocating Engines

The following nomenclature is used to describe reciprocating engines for driving compressors:

- 2-Stroke—Two piston strokes required to produce one power stroke per cycle
- 4-Stroke—Four piston strokes required to produce one power stroke per cycle
- Integral Engine Compressor—Gas compressors attached to engine crankshaft throws
- Naturally-Aspirated—Combustion air induced by the power cylinders
- Scavenged Air Supply—Combustion air supplied by a powered blower or compressor
- Turbocharged Air Supply—Exhaust gas-operated turbine drives centrifugal air compressor
- High Speed—Above 800 r/min
- Medium Speed—400-800 r/min
- Low Speed—Below 400 r/min
- Heavy Duty—Designed to carry rated load continuously
- Lean Burn—Designed to produce minimum detrimental exhaust emissions

*Courtesy of Energy Services Group of Cooper Industries*

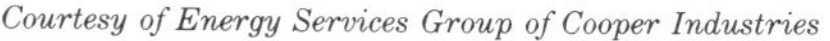

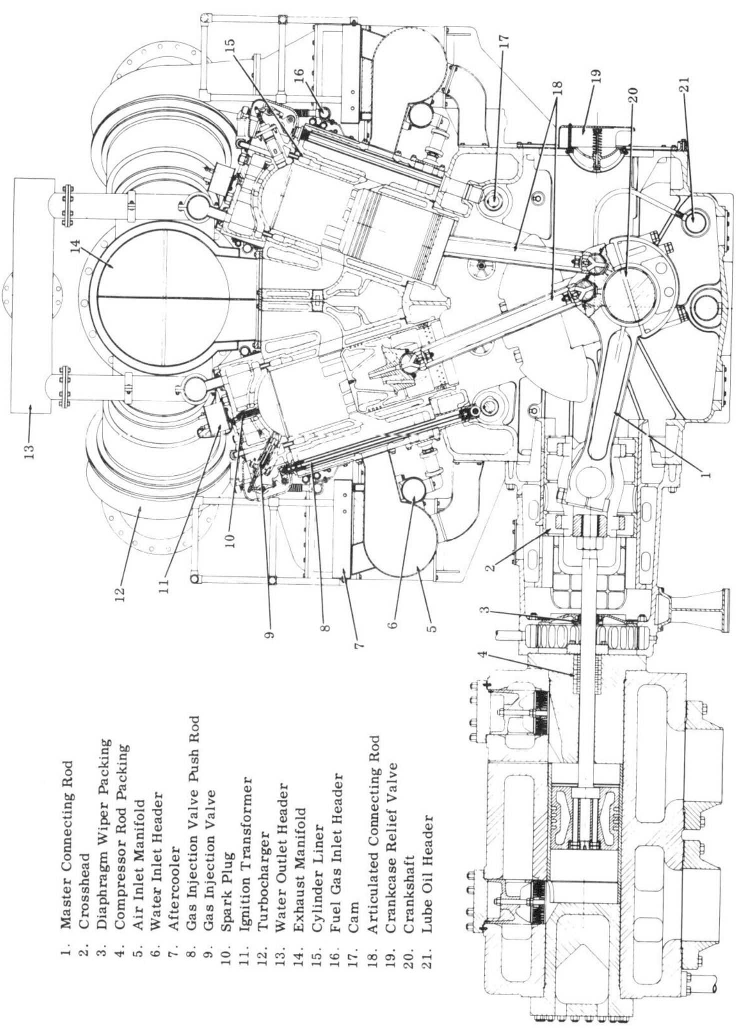


***Figure 7.* Cross-section of a 2-stroke natural gas engine with integral compressors**

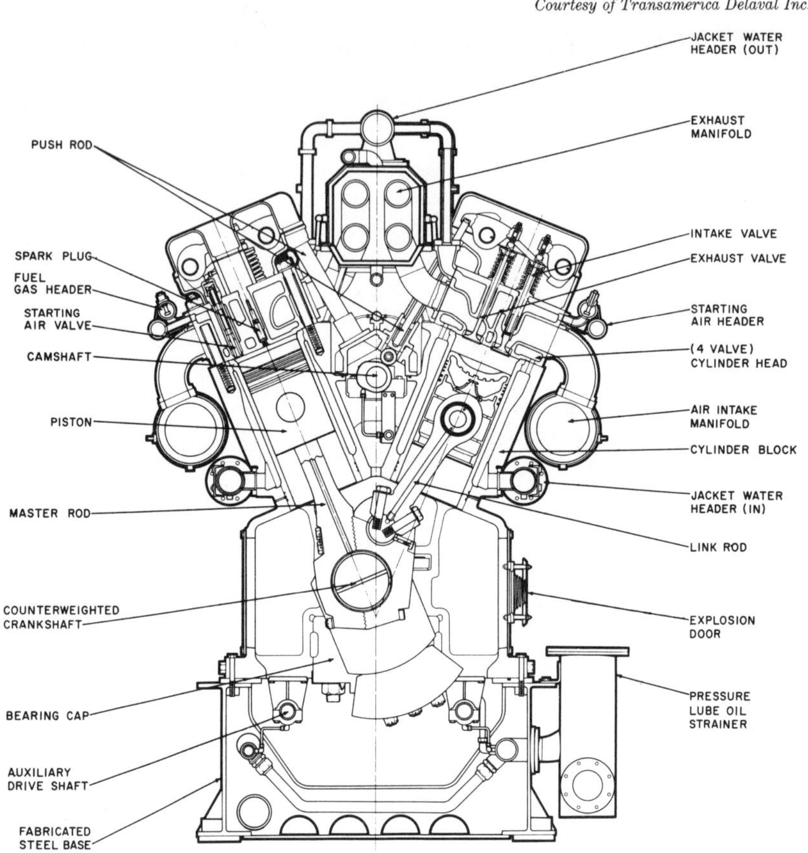


*Figure 8.* **Cross-section of 4-stroke natural gas engine**

- Opposed Piston – Power pistons operate against each other
- Horizontal – Power pistons operate in a horizonal direction
- Vertical – Power pistons operate in a vertical direction.

Reciprocating units are available in many combinations of size, mechanical design, and operating principles.

## Reciprocating Engine Size Criteria

Reciprocating gas engines are available in sizes ranging from 20 hp (15 kW) to several thousand horsepower. The largest unit now available with an integral compressor is 13 500 hp (10 070 kW). The size of a unit is determined by the developed mean effective pressure (MEP), engine speed, and design. A 2-stroke engine will develop about 50 percent more power per cylinder than a 4-stroke engine. Increased output power is achieved by mechanically providing scavenging air which is essential for 2-stroke engines, and often is applied to 4-stroke engines. The most popular method of providing scavenging air is by means of an engine exhaust-driven turbocharger. Normally gas engines are designed to operate within a speed range of 200 to 3 600 r/min. A small integral engine/compressor unit is shown in Figure 9.

## Reciprocating Engine Installation Factors

The selection of a high-speed, high-power unit usually is dictated by weight and size. To facilitate installation, many of these units can be packaged with other components and mounted on structural steel skids.

If space is not restricted, ample room should be provided to facilitate operations and maintenance. If possible, a building should be provided with adequate work space and an overhead crane to transfer parts.

## Reciprocating Engine Operating Characteristics

Reciprocating engines readily are adapted to remote automated operations. However, adequate safety shutdown devices should be provided on units that are unattended. Two ways of controlling the power loading are: maintain the torque and vary the speed to achieve the desired compressor discharge pressure, and vary the torque and maintain the engine speed. Maintaining rated torque and varying the engine speed within the limits of the unit usually results in better efficiency.

Reciprocating engines impose dynamic unbalanced forces that are acceptable in conventional engine installations but are not acceptable on offshore gas production platforms. Specially balanced engines for critical applications can be supplied by engine manufacturers.

## Reciprocating Engine Efficiency

Since thermal efficiency decreases rapidly as the load is reduced, the unit should be operated at the rated power for maximum efficiency. Although reciprocating engines are available with thermal efficiencies as high as 40 percent when operated at the rated load, most units normally operate in a range of 25 to 40 percent. A typical heat balance diagram for a 2-stroke reciprocating engine in a compressor station operation is shown in Figure 10.

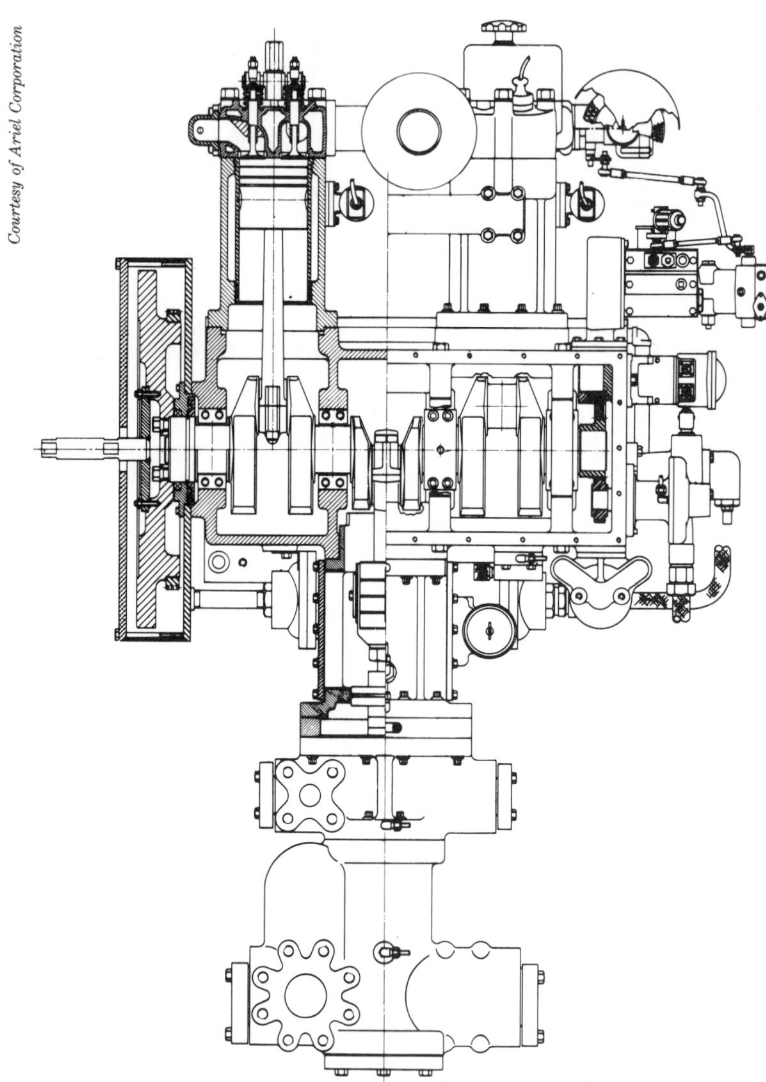


*Figure 9.* Top cross-section of a small reciprocating integral engine/compressor

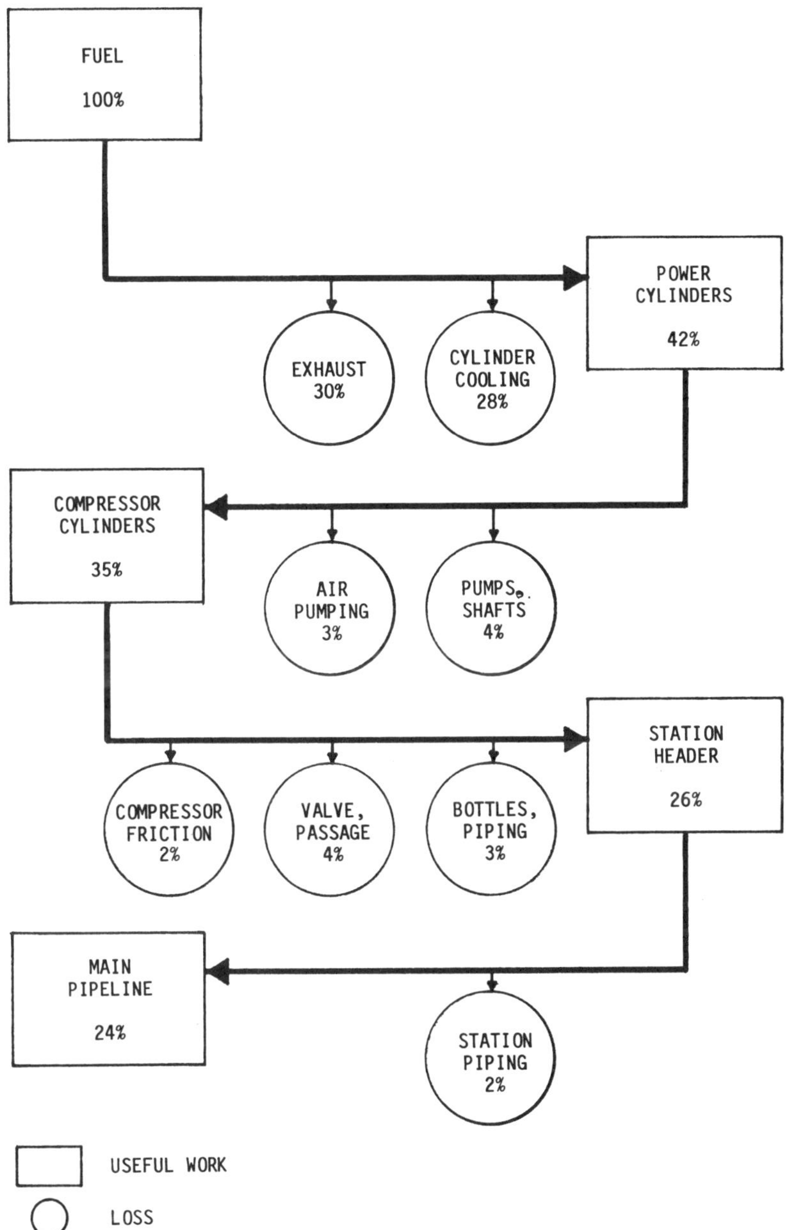


*Figure 10.* Typical heat balance diagram for 2-stroke reciprocating engine

### Exhaust Emissions from Reciprocating Engines

Emissions from reciprocating engine exhausts depend on the composition of the fuel. Natural gas is a clean fuel usually containing only a minimal amount of sulfur, particulates, and other harmful substances.

The major problem encountered in engine exhaust pollution is caused by oxides of nitrogen ($NO_X$). $NO_X$ is produced by high temperatures and pressures generated in the power cylinders. Consequently, the concentration of $NO_X$ increases proportionally with the efficiency of a unit. At present, there are no standards limiting the quantity of emissions that may be discharged, but ambient air standards limit the amount of nitrogen dioxide ($NO_2$) present at ground level.

The $NO_X$ concentration in the ambient air is reduced by increasing the stack flue gas velocity and the height of the stack. This disperses the emissions over a larger area thereby reducing the concentration. Research is being conducted to find ways of reducing $NO_X$ exhaust. Two methods currently under consideration are catalytic converters and recycling a portion of the exhaust gases.

## ROTARY ENGINES

One engine manufacturer offers a rotary engine utilizing the Wankel configuration for gas compression applications. This unit combines many features of the reciprocating engine and the gas turbine.

The gas operated unit is offered in two sizes, 550 and 1 100 hp (410 and 820 kW). The speed range is from 600 to 1 200 r/min, and the unit has a thermal efficiency of 32 percent at the rated load and speed. The unit is compact, operates with little vibration, and lends itself to packaging in a compact, skid-mounted engine/compressor unit.

## GAS TURBINES

Gas turbines are available in many different makes and models; however, the operating principles of all models are similar. Basically a gas turbine consists of an air compressor, combustor, and expander turbine or turbines. Ambient air is drawn through a filter and silencer into the compressor where it is compressed to 5 or 6 atmospheres. The compressed air is then discharged into a combustion system to support the fuel combustion. The resulting combustion products and any excess air that is present are directed by the first-stage nozzle into the first-stage turbine rotor blades. Gas from the first-stage rotor is directed in succession to the additional turbine stages, and finally exhausted to the atmosphere or to a waste heat regeneration unit which has been installed to utilize exhaust heat energy.

If the turbine is a single-shaft design, the compressor, all turbine stages, and the driven equipment rotate at the same speed.

Two-shaft turbines are better suited for pipeline compressor drive service. A two-shaft gas turbine is shown in Figure 11. The air compressor is powered by the first stage or the stages making up the first-stage rotor, usually referred to as the gas generator, gas producer, or high-pressure turbine. One or more turbine stages drive the second shaft, which is coupled to the load. This section is known as the power turbine, free turbine, load turbine, or low-pressure turbine. The maximum allowable firing temperature has been increased from about 1 350°F (730°C) in earlier turbine designs to 2 000°F (1 090°C) in present day turbines. This temperature increase has been possible because of metallurgical advances and improved cooling techniques.

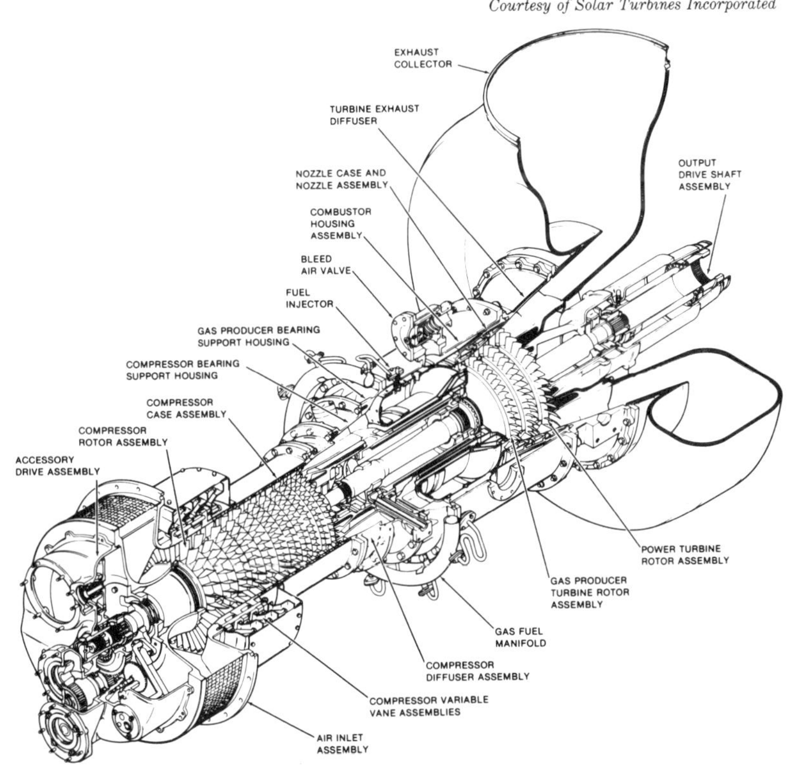


*Figure 11.* **Cutaway of two-shaft gas turbine engine**

### Types of Gas Turbines

The following nomenclature is used to describe the types of gas turbine configurations and operation:
- Single-Shaft: A design in which a single rotor serves to couple the combustion air compressor, turbine wheel, and the power takeoff – making them run at the same speed
- Two-Shaft: A design in which the combustion air compressor and compressor turbine are on one shaft forming a gas generator, and the power turbine is on a separate shaft for the power takeoff. This arrangement allows each shaft to run at different speeds and allows a greater turndown ratio at the output shaft
- Simple Cycle: Exhausted directly to atmosphere with no heat recovery
- Regenerative Cycle: Exhaust gas flows through a regenerator or heat exchanger, and increases the temperature of compressor discharge air prior to combustion
- Combined Cycle: Exhaust gas is used to generate steam for separate steam turbine
- Aircraft Derivative: A two-shaft design in which the jet aircraft engine, after removal of the nozzle, becomes the gas generator that drives a separate power turbine.

### Gas Turbine Size Criteria

Gas turbines are available in sizes ranging from 1 000 to 30 000 hp (745 to 22 370 kW) in both aircraft derivative and industrial types.

### Gas Turbine Installation Factors

Aircraft derivative turbines are lightweight with a high power output. Because of their size they are easily packaged with a gas compressor on a structural steel skid, requiring a small foundation.

Industrial gas turbines are still relatively lightweight considering their power output. Foundation requirements are not severe, and the units are easy to automate. A typical industrial gas turbine-driven centrifugal compressor package is shown in Figure 12.

### Gas Turbine Operating Characteristics

International Organization for Standardization (ISO) documents rate the gas turbines at 15°C (59°F) air inlet temperature at sea level. At lower ambient temperatures the power output capability increases con-

*Courtesy of Solar Turbines Incorporated*

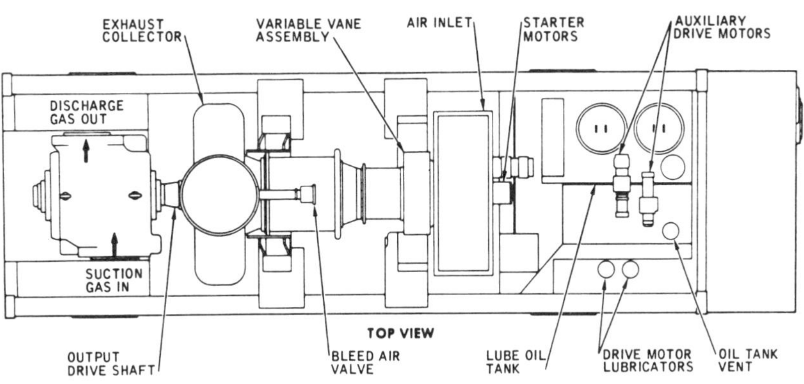


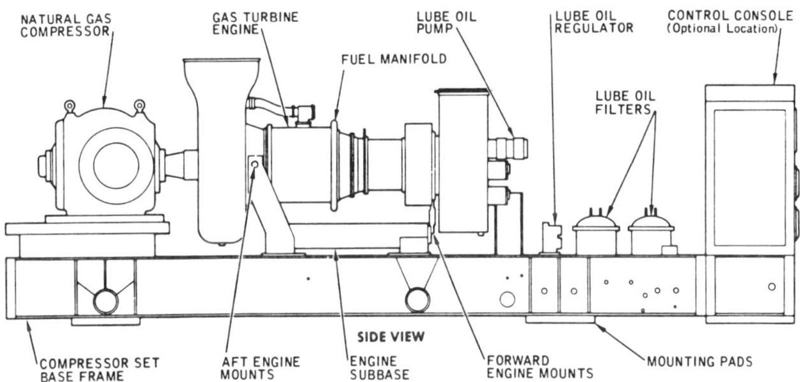


*Figure 12.* **Turbine-driven centrifugal compressor package**

siderably, and as the barometric pressure decreases at high elevations, the power output diminishes. Turbines are rated at various loads and operating speed ranges. Driven-equipment rated speed must be designed to match the turbine speed; therefore, a speed increaser or reducer may be required.

## Gas Turbine Efficiency

Thermal efficiencies for gas turbines range from about 18 to 28 percent, depending upon the make and model. Heat recovery equipment, such as steam generators and regenerators, improves efficiency by lowering the fuel input requirements.

## Exhaust Emissions from Gas Turbines

Exhaust emissions from gas turbines are less concentrated than those from reciprocating engines. The turbine section operates at a sufficiently high temperature to ensure complete combustion of the fuel, while excess air reduces the concentration of the combustion products. Exhaust stacks are usually tall enough so that the gas leaving the stack at a high velocity is dispersed into the atmosphere over a wide area. Water injection and redesigned combustion chambers are used to minimize $NO_X$ emissions.

Noise from gas turbines can be controlled with intake and exhaust silencers and acoustical dampers.

## STEAM TURBINES

Steam turbines driven by fired boilers have had limited application in the gas industry. As energy has become more costly, much interest is being directed toward the use of turbines driven by unfired boilers, utilizing waste energy. Waste heat recovery can improve the efficiency of a plant and provide increased gas compressor capacity.

Two factors are important when a steam application is being considered – the water supply and the mode of operation. An adequate water supply must be available to provide feed water for the boilers and cooling water for the condensers. The quality of the water supply is also important. In many cases the make-up water must be demineralized before it enters the system.

Cooling water from a stream, pumped through a shell-and-tube condenser, is usually the least expensive way to condense the steam, assuming that the water does not contain biological life that can clog the condenser. Another way to provide cooling is by using an atmospheric cooling tower. Air-cooled condensers can be used but are subject to freezing in low-temperature applications. The mode of operation is important because in most cases steam plants are manned during operation.

## Types of Steam Turbines

A multistage reaction turbine usually is selected to drive centrifugal compressors. Depending upon the size of the application, a combination impulse/reaction turbine can be applied.

## Steam Turbine Size Criteria

The size of the steam turbine will be dictated by the amount of exhaust energy available and the power required for the application. A steam system utilizing the exhaust from reciprocating engines can recover additional energy equivalent to approximately 10 percent of the power generated by the reciprocating engines; thus, a 10 000-hp (7 460-kW) reciprocating engine installation will drive a 1 000-hp (746-kW) steam-turbine-driven compressor. With a gas turbine, which produces more high-energy exhaust, approximately 40 percent or more of the nameplate power of the original turbine can be recovered.

## Steam Turbine Installation Factors

A steam plant is larger than any other type of equipment presently used to compress gas. It occupies more space and contains more metal surface. The waste heat recovery boilers required to supply steam to the turbine may be receiving energy from the exhaust of 10 to 15 reciprocating units. A steam plant is impractical in any application where space is limited, and cycle efficiency and fuel economy are not important factors.

## Steam Turbine Operating Characteristics

When a waste heat recovery boiler driving a steam turbine is operated in conjunction with the original prime mover, the gross power of the system, when plotted on a graph, tends to flatten out—that is, the system power would be nearly the same throughout the year.

## Steam Turbine Efficiency

Combined cycles using exhaust heat from a gas turbine to provide steam for the steam turbine can increase the overall thermal efficiency by 10 to 15 percent. Studies indicate that the efficiency of a system utilizing an older, simple-cycle gas turbine may be increased from 22 to 36 percent. System efficiencies of over 50 percent can be expected using the most modern turbines.

## Exhaust Emissions from Steam Turbines

No additional exhaust emissions are generated by steam turbines which utilize as their only source of energy the exhaust gas from another prime mover. The exhaust gases will be cooler; therefore, the dispersion of the gases to the atmosphere will not be as effective. A means of disposal of the boiler blowdown and chemicals from the feed water treatment must be provided when considering a steam plant. If a water-cooled condenser is used, thermal pollution may become an important consideration in the future.

## ELECTRIC MOTORS

Electric motor prime movers for compressor stations have advantages in overall construction cost, ease of control, minimum pollution due to emissions, small building requirements, and lower maintenance costs. Generally a major disadvantage is the relatively high cost of electric power. Another is that any momentary interruption in the power supply may shut down the station. In general other types of prime movers are less expensive and more dependable.

The principal consideration for a proposed compressor station with electric motor-driven compressors is the location. It must be located where electric power is available with sufficient capacity at a relatively low cost. Each site should be evaluated separately.

It is desirable to have an electric power supply from two independent power sources with automatic switching from one source to the other in case one fails. Electric motors may be used to drive either reciprocating or centrifugal compressors, and are available in different speed ratings for either application.

### Applications for Electric Motors

Synchronous motors are the best type to use for the following reasons:
- Higher efficiency
- Ability to provide reactive power to other equipment, resulting in lower power cost
- Adaptability to large frames
- Low kilovolt-ampere inrush requirements during start-up.

Electric prime mover compressors usually are installed inside a building but may be installed outside. In either case the installation must comply with the National Electrical Code (NEC) and the Occupational Safety and Health Administration (OSHA) requirements.

In some cases use of explosion-proof apparatus may be advisable. However, since larger motors are not available in explosion-proof housings, totally enclosed, forced ventilation motors are used. An outside source of clean air is forced through the enclosure under a positive pressure. If the positive air pressure is lost by failure of a fan, or for any other reason, the motor must be designed to stop automatically.

## Electric Motor Size Criteria

Electric motors range in size from 25 to 15 000 hp (19 to 11 200 kW) with speeds from 180 to 1 800 r/min and voltages from 460 to 13 200 volts. The motor speed must match the type and speed of the compressor being considered.

Reciprocating compressors are driven in the lower speed ranges and may be connected to a motor by a mechanical or magnetic coupling. Centrifugal compressors rotate at higher speeds. For speeds above 1 800 r/min, a gear box may be required.

It is imperative that the motor and compressor speed, starting torque, and inertia of the two components be compatible.

## Electric Motor Installation Factors

In addition to having a reliable source of power at low rates, other factors that must be considered include:
- Substation with transformers and switching
- Switchgear and controls
- Station operating demand factor.

Generally a power company will not supply or maintain the power transformers and substation switching equipment for the compressor station. This is usually the owner's responsibility. Switchgear includes the main breakers, contactors, and field application panels for each prime mover as well as distribution equipment for all loads such as vent fans, pumps, and lighting. The equipment must be remote from hazardous prime mover areas and should be located inside a building for weather protection.

Synchronous motors require an external excitation source, either a static exciter or a motor/generator set. Some of the larger synchronous motors are started as induction motors. At about 97 or 98 percent rated speed, excitation is applied at the proper instant for the motor to pull into step, and from then the motor runs as a synchronous motor. Motors may be started at full or reduced voltage depending upon the power company's requirements and the frequency of starting.

Basic controls often use direct current with a battery backup to operate protective devices if and when the prime mover source is lost. The following are other control factors that should be considered:
- Connect the motor starter winding to the line at full or reduced voltage, as required
- Provide coordinated, instantaneous overcurrent protection from short circuits or other fault conditions

- Provide thermal-type delayed overcurrent protection from abnormal load conditions
- Apply field excitation at proper "pull-in speed" and rotor angle to sychronize the motor
- Protect the rotor cage winding upon loss of excitation or synchronization
- Protect the motor against extreme voltage dips or loss of voltages. This protection may be instantaneous, or it may be time-delayed to restart the motor if the voltage returns to normal within about 2 seconds
- Indicate motor input by instrumentation
- Consider including field current adjustment to control power factor
- Provide safety for operating personnel
- Consider including added protective relays for differential, ground fault, reverse phase, underfrequency, etc.
- Remote start/stop control on-site or by supervisory control at any distance.

## Electric Motor Operating Characteristics

One of the disadvantages of electric prime movers is the lack of speed control. The motors run at constant speed, not variable as is possible with a reciprocating engine or turbine. Because of the high minimum monthly charge by power companies, electric prime movers are more economical when the station is operated at least on a monthly basis. Electric motor-driven compressors can seldom be justified for peaking operations.

One advantage of electric motor-driven compressors is that maintenance is low compared to that for turbines or reciprocating engines.

## Electric Motor Efficiency

The efficiency of synchronous motors increases with higher power ratings, varying between 89 to 96 percent at the rated load. The unit efficiency will vary depending on the type of coupling used. Magnetic couplings and gear boxes will decrease the overall efficiency slightly.

## Emissions from Electric Motors

Both noise levels and emissions from electric motor-driven prime movers are extremely low as compared to turbines or reciprocating engines.

# GAS COMPRESSORS

The following describes the basic types and characteristics of gas compressors and how they are applied to the pipeline system, as well as methods of determining size, capacity, and power requirements.

## TYPES OF COMPRESSORS

While gas compressors are available in a variety of types, sizes, and physical configurations, they are all designed for the same purpose. In the natural gas industry the compressor is used to restore pipeline energy losses incurred in the transportation of gas, or to convey gas from a low-pressure source to a higher pressure receiver.

The two basic types of compressors currently in use in natural gas service are the reciprocating compressor and the rotating compressor. All reciprocating compressors are positive displacement machines. Rotating compressors can be either positive displacement or dynamic. In a positive displacement machine the compression is accomplished by confining a discrete quantity of gas within an enclosure and then increasing its pressure by reducing its volume. In a dynamic machine the pressure increase is achieved through a velocity change in a continuous-flow stream rather than a volume reduction of an isolated quantity of gas.

### Reciprocating Compressors

Reciprocating compressors utilize the motion of a piston in a cylinder to compress the gas. The piston's motion is generated from a connecting rod attached to a crankshaft. One or several compressor cylinders may be attached to a common crankshaft. If the compressor and the prime mover share the same crankshaft, the compressor unit is termed an integral unit. If the compressor is external to the prime mover and is driven from the output shaft, a separable unit is formed. The prime mover for a separable compressor can be an internal combustion engine, a gas turbine, a steam turbine, an electric motor, or a turbo-expander. Figures 13 and 14 show typical integral and separable reciprocating units. A cross-sectional view of an integral engine/compressor is shown in Figure 7. A cutaway view of a separable compressor is illustrated in Figure 15.

### Rotating Compressors

Rotating compressors utilize a rotating element to compress the gas.

*Courtesy of Dresser Industries, Dresser Clark Division*

**Figure 13.** Reciprocating, integral engine/compressor unit

*Courtesy of Cooper Energy Services, Superior Division*

**Figure 14.** Skid-mounted, reciprocating, separable engine/compressor package

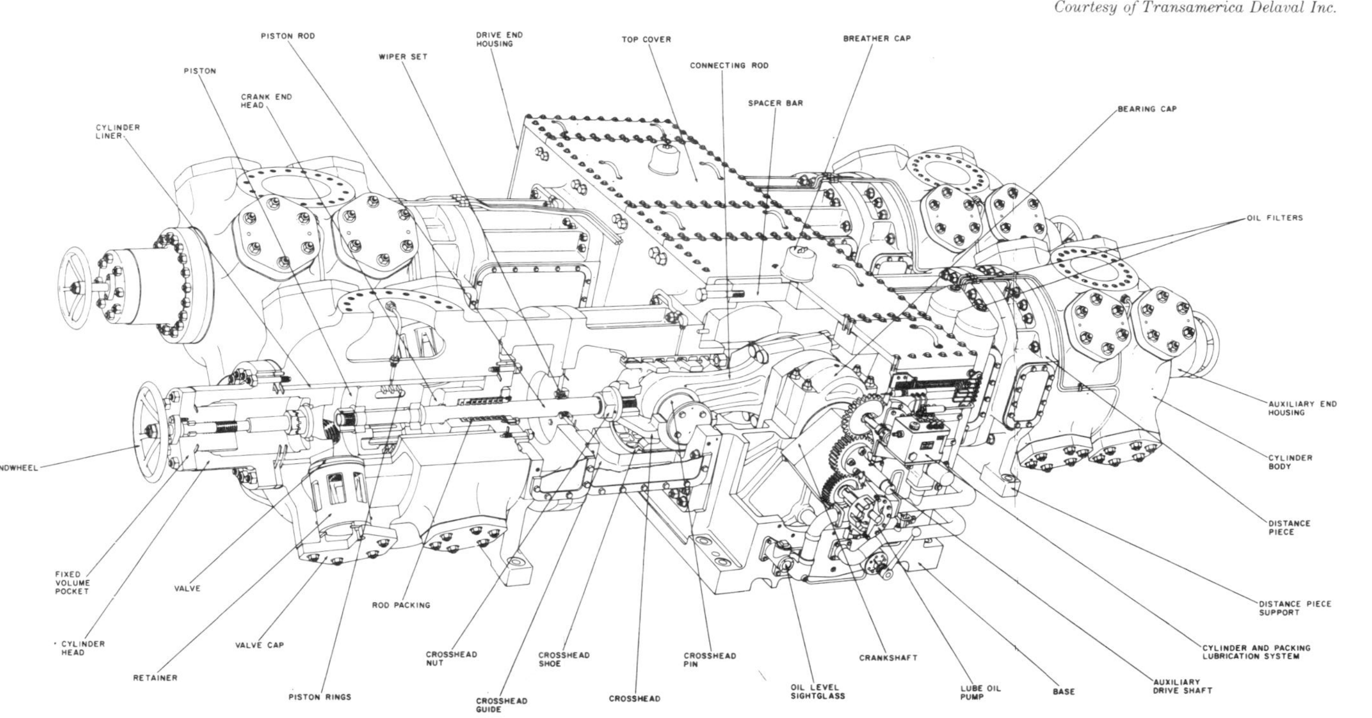


*Figure 15.* **Cross-section of a reciprocating separable compressor**

The rotating element commonly is referred to as a wheel, impeller, or rotor depending on the type of machine and the manufacturer's nomenclature. The rotating member is contained within a pressure-tight housing, or compressor case, and is driven from the output shaft of a prime mover located adjacent to the compressor case. The prime mover can be an internal combustion engine, gas turbine, steam turbine, electric motor, or turbo-expander.

A positive-displacement rotating compressor utilizes sliding vanes or intermeshing rotors to trap a quantity of gas and reduce its volume.

When the pressure increase is achieved through a velocity change in a continuous flow stream rather than a volume change of an isolated quantity of gas within the machine, the process is known as dynamic rotating. Dynamic rotating compressors utilize a rapidly spinning wheel, or impeller, to impart a higher velocity to a steady flow of gas. The high velocity gas leaving the impeller passes through a stationary diffuser section which converts velocity energy to pressure energy. Because of its wide usage in the gas industry, a dynamic machine known as a centrifugal compressor will be discussed further under Dynamic Compressor Calculations. Figures 16 and 17 show typical positive displacement and dynamic rotating compressors.

Other types of rotating compressors utilized or considered by the gas industry for specific or unique applications are:

- Axial Flow Compressor—The axial flow compressor moves and compresses gas in a straight-through path parallel to the rotor shaft. A pressure rise is achieved by the interaction of a series of alternate stationary and rotating turbine-type blades. The axial flow compressors are not suitable for service on gas with fouling tendencies; therefore, the most frequent application is to compress filtered air in the gas generator section of a gas turbine machine.
- Mixed Flow Compressor—The mixed flow compressor combines the characteristics of the centrifugal and axial flow compressors.
- Liquid Piston Compressor—A liquid piston compressor is a rotary positive displacement machine in which water or another liquid is used to compress the gas.

## COMPRESSOR APPLICATIONS

The proper application of gas compressor machinery requires many considerations involving the characteristics of the intended service, as well as the characteristics of a particular type of compressor.

The terms and nomenclature commonly used in compressor application provide the necessary background and basis for the thermodynamic

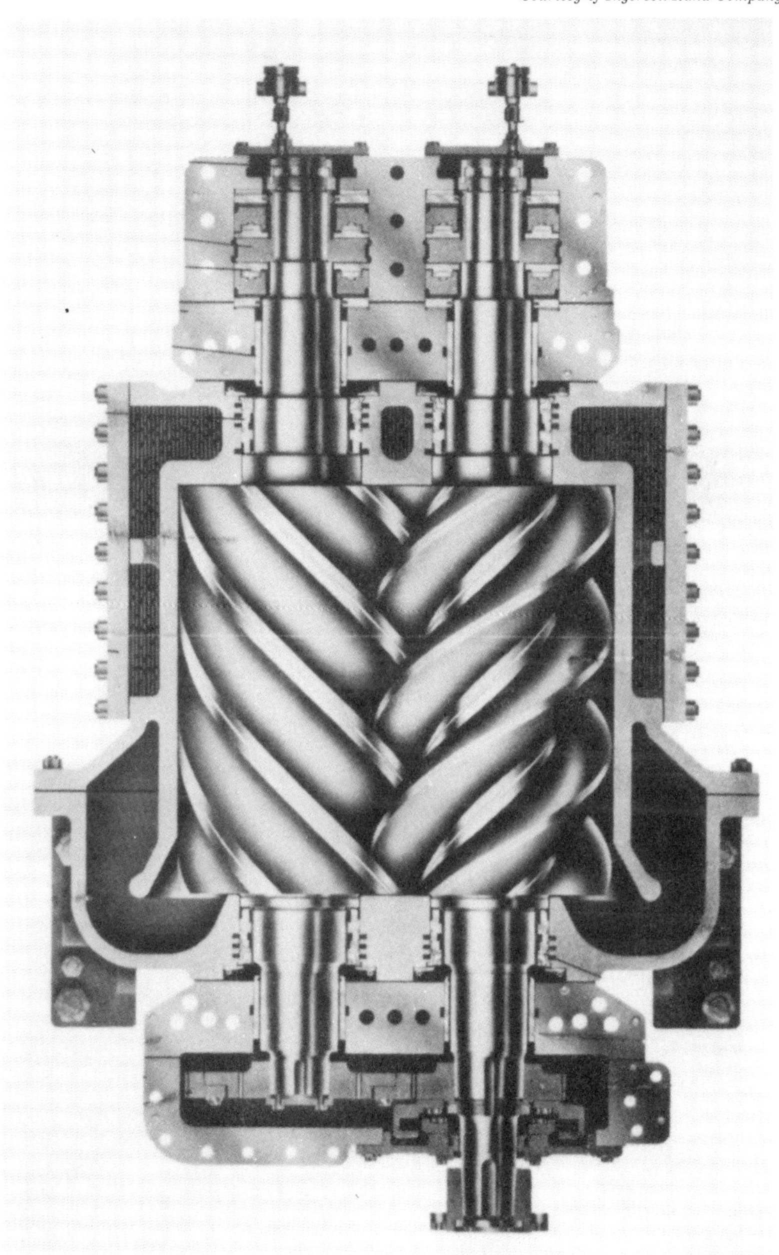

*Figure 16.* Positive displacement rotary compressor

Courtesy of Ingersoll-Rand Company

*Figure 17. Centrifugal compressor unit*

relationships and formulas described below under Compressor Calculations.

## Basic Terms and Definitions

The following terms apply to all compressors regardless of type:

- Suction Pressure–The gas pressure existing on the suction, or intake, side of the compressor unit expressed in lbf/in² (kPa) gauge or lbf/in² (kPa) absolute. The exact point of measurement of suction pressure is important in the final sizing and selection of the compressor
- Discharge Pressure–The gas pressure existing on the discharge, or outlet, side of the compressor unit expressed in lbf/in² (kPa) gauge or lbf/in² (kPa) absolute; as with suction pressure, the exact point of measurement is important in the final sizing and selection of the compressor
- Compression Ratio–The quotient obtained from dividing the absolute discharge pressure by the absolute suction pressure
- Capacity–The volumetric quantity of gas expressed in MMcf/d (m³/d) or MMscf/d [m³/d (st)] that passes through the compressor unit; capacity is basically a function of compressor geometry, gas composition, temperature, pressure, and compressor speed
- Horsepower–The time rate of work that is developed by the prime mover as required by the compressor; the total required horsepower is a function of capacity, compression ratio and compressor (thermodynamic) efficiency, and mechanical efficiencies of the power transmission system.

## Service Considerations

The intended service under which a compressor unit is required to operate is the first item that must be defined. The following three basic types of compressor services are those most often found in natural gas pipeline operations.

### Transmission Service

A compressor in transmission or relay service raises the pressure in a transmission pipeline system to overcome the effect of frictional losses in the pipeline itself, and to maintain the required pressure at the various delivery points.

Transmission compressors may be of any horsepower and size depending upon the quantity of gas being transported and the pressure levels required. A typical transmission or relay station operates at low to moderate compression ratios (1.1 to 2.5). The operation of the transmission station is highly dependent upon market or customer requirements and may be subject to large capacity variations. Gas storage or peak-shaving facilities can be utilized to stabilize the variations.

### Storage Service

A compressor in storage service pumps gas from a low-pressure storage field to a higher pressure transmission line and/or pump gas from a low-pressure transmission line to a higher pressure storage field.

Storage compressors must be flexible enough to allow operation across a wide band of suction and discharge pressures and volume variations. Gas injection, or pump-in service, may require compression ratios from 2 to 5 depending upon the geological properties of the storage field and the available suction pressure. During conditions of high compression ratios the injection volume may be quite low. Gas withdrawal, or pump-out service, may require compression ratios of 1.1 to 3 at correspondingly higher volumes due to peak requirements of the markets.

### Production Service

A compressor in production, or gas-gathering, service pumps gas from low-pressure wells or gathering systems to a higher pressure gathering or transmission line.

Generally production compressors are exposed to relatively constant discharge pressures and declining suction as gas reserves are depleted. The compressor capacity must be matched to the deliverability of the production field. Contrasted to transmission or storage compressors, the service life of these units may be shorter because of the speculative nature of natural gas production. Production compression ratios range from moderate to high (2 to 30).


## COMPRESSOR CALCULATIONS

This discussion provides the mathematical relationships necessary to determine the size of a compressor, the power required to compress a given volume of gas, and mechanical limits. Because of the many published references on this subject, an abbreviated approach will be taken with regard to the thermodynamic and aerodynamic theory.


### Reciprocating Compressor Calculations

As mentioned earlier, a reciprocating compressor is a positive displacement machine that elevates the pressure of a trapped volume of gas by the action of a piston in a cylinder. In each cylinder, valves are provided that operate automatically by pressure differential (like check valves) to admit and discharge the gas. The inlet valve opens when movement of the piston has lowered the cylinder pressure to below inlet line pressure. The discharge valve closes once the pressure in the cylinder drops below the pressure in the discharge line, thus preventing flow reversal. Figure 18 gives an exploded view of a typical compressor plate valve for a reciprocating compressor.

All reciprocating compressors have a space between the end of the piston and the cylinder head. The volume formed by this space plus the volume formed by the valve cavities is called the fixed clearance. This clearance, plus any variable clearance (see Compressor Characteristics and Load/Capacity Control), constitutes the total clearance volume.

### Basic Cycle of Reciprocating Compressors

Figure 19 illustrates an ideal compression cycle on a pressure-volume, or $p$-$V$, diagram. The events during this cycle are as follows:

- Compression ($a$–$b$)—With the suction valve in the closed position and the cylinder filled with gas at suction pressure, the piston is in position $a$. As the piston moves from $a$ to $b$, gas is compressed isentropically (no heat transferred in or out of the gas, and no frictional losses) until the pressure in the cylinder reaches the discharge pressure.
- Discharge ($b$–$c$)—At point b the discharge valve opens and remains open until the piston reaches the end of the stroke at point $c$.
- Expansion ($c$–$d$)—With the discharge valve in the closed position, the compressed gas trapped in the clearance volume expands isentropically from discharge pressure to suction pressure as the piston moves from $c$ to $d$.
- Suction ($d$–$a$)—At point d the suction valve opens and permits gas at suction pressure to enter the cylinder as the piston moves from $d$ to $a$.

### Reciprocating Compressor Performance

To determine the quantity of gas that a specific compressor can pump, the displacement, volume of gas at suction conditions, and suction volumetric efficiency must be known.

**Displacement.** The displacement of a compressor normally is expressed in cubic feet per minute (cubic metres per second), or ft³/min (m³/s). This is a function of the area of the piston face, the length of the stroke, and the number of strokes per unit of time. Figure 19 shows the cycle as related to a single-acting compressor; that is, compression occurs only on one end of the piston. Most reciprocating compressors

*Courtesy of Hoerbiger Corporation*

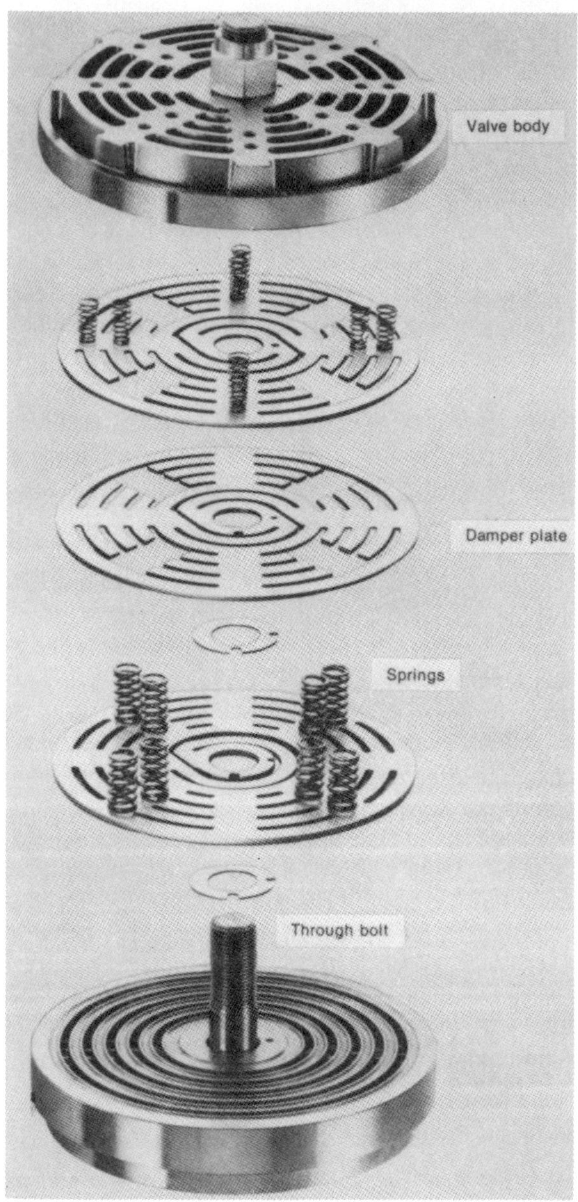


*Figure 18.* Exploded view of a high-speed compressor plate valve

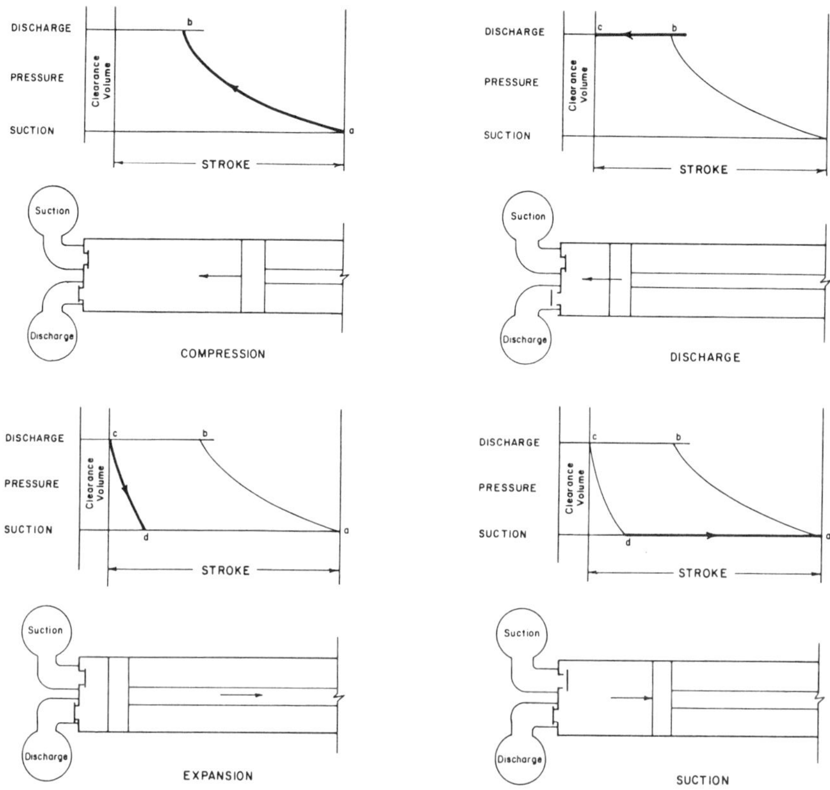


*Figure 19.* **P-V diagram illustrating ideal reciprocating compressor cycle**

in the gas industry are double-acting. Gas is compressed alternately on each end of the piston. Because of the cylinder area occupied by the piston rod, the available area for compression on the crank end of the piston is less than that on the head end of the piston.

## Single-Acting Compressor

$$PD = \frac{A_{\mathrm{H}} \times l \times N}{1\ 728} \ \mathrm{ft^3/min}$$
Equation 1a

or

$$PD = \frac{A_{\mathrm{H}} \times l \times N}{6 \times 10^9} \ \mathrm{m^3/s}$$
Equation 1b

Where: $PD$ = Piston displacement in desired units, ft³/min or m³/s
    $A_H$ = Area of head end of the piston, in² (mm²)
    $l$ = Length of stroke, in (mm)
    $N$ = Compressor rotation speed, r/min

### Double-Acting Compressor

$$PD = \frac{(A_H + A_C) \times l \times N}{1\ 728} \text{ ft}^3/\text{min} \qquad \text{Equation 2a}$$

or

$$PD = \frac{(A_H + A_C) \times l \times N}{6 \times 10^9} \text{ m}^3/\text{s} \qquad \text{Equation 2b}$$

Where: $A_C$ = Area of crank end of the piston, in² (mm²)

Most compressor manufacturers publish data listing displacement at full compressor speed for each size (bore, stroke, rod size) cylinder that is manufactured.

**Volume.** Because the gas entering the compressor cylinder is normally at a pressure higher than atmospheric, the volume of gas displaced will be greater than the compressor displacement. Therefore, by Boyle's Law it is necessary to multiply the displacement by the ratio of the actual to the base suction pressure in lbf/in² (kPa) absolute. To illustrate the calculations in this text, the base suction pressure has been chosen as 14.4 lbf/in² (99 kPa) absolute. The appropriate factors should be used if another pressure base is required. Similarly, a base suction gas temperature must be selected. By Charles' Law it will be necessary to multiply the displaced volume by the ratio of the absolute base temperature to the absolute suction temperature, in °R (K). Inlet volume to any stage of the compressor may be calculated from the standard volume data by using the proper inlet pressure and temperature, correcting for compressibility and moisture content where necessary.

**Volumetric Efficiency.** Volumetric efficiency factor ($E_V$) in a reciprocating compressor is the ratio (percent) of the actual delivered volume flow rate (at inlet conditions) to the piston displacement. If there was no clearance volume to re-expand and delay the opening of the suction valve, and if there was no pressure drop or heat transfer into the gas, the cylinder could deliver 100 percent of the entire piston displacement into the line. The effect of clearance volume on volumetric efficiency depends on the compression ratio, and a physical property of the gas known as the ratio of specific heat capacities, or $k$ value. This constant may be calculated from a gas analysis or found in reference tables in a general form. Typical $k$ values for natural gas vary from 1.26 to

1.31. The theoretical volumetric efficiency factor may be calculated as follows:

$$E_V = 100 - Cl \times \left[ \left( \frac{p_\mathrm{d}}{p_\mathrm{s}} \right)^{\frac{1}{k}} - 1 \right]$$

Equation 3

Where: $E_V$ = Volumetric efficiency factor, percentage
$Cl$ = Cylinder clearance expressed as a percentage of piston displacement
$p_\mathrm{d}$ = Pressure at the discharge flange, lbf/in² (kPa) absolute
$p_\mathrm{s}$ = Pressure at the suction flange, lbf/in² (kPa) absolute
$k$ = Ratio of specific heat capacities, $\dfrac{c_p}{c_V}$

Equation 3 is based on an ideal $pV$ diagram, and does not take into account factors that affect the volumetric efficiency under actual operating conditions. To compensate for actual conditions, a volumetric efficiency loss correction factor ($L$) is introduced which accounts for the inlet gas preheating, pressure drop across the valves, internal leakage, gas friction, etc. While most of these items are inlet pressure and pressure ratio dependent, non-lubricated compressor construction will double the $L$ value. Although this value must be determined empirically, in natural gas compression the numerical value of the compression ratio is a fair approximation of $L$ expressed as a percentage. The generalized equation is:

$$E_V = 100 - L - Cl \times \left[ \left( \frac{p_\mathrm{d}}{p_\mathrm{s}} \right)^{\frac{1}{k}} - 1 \right]$$

Equation 4

Where: $L$ = Volumetric efficiency correction factor, percent

If the compressibility factor $Z$ at discharge conditions differs significantly from that at the inlet, better results will be obtained by multiplying the term $(p_\mathrm{d}/p_\mathrm{s})^{1/k}$ by the ratio $Z_\mathrm{s}/Z_\mathrm{d}$

Where: $Z_\mathrm{s}$ = Compressibility factor at suction conditions
$Z_\mathrm{d}$ = Compressibility factor at discharge conditions

Figure 20 shows the relationship between the volumetric efficiency factor, compression ratio, and clearance – assuming $L$ is equal to the absolute value of $p_\mathrm{d}/p_\mathrm{s}$ – for a gas having a $k$ value of 1.3.

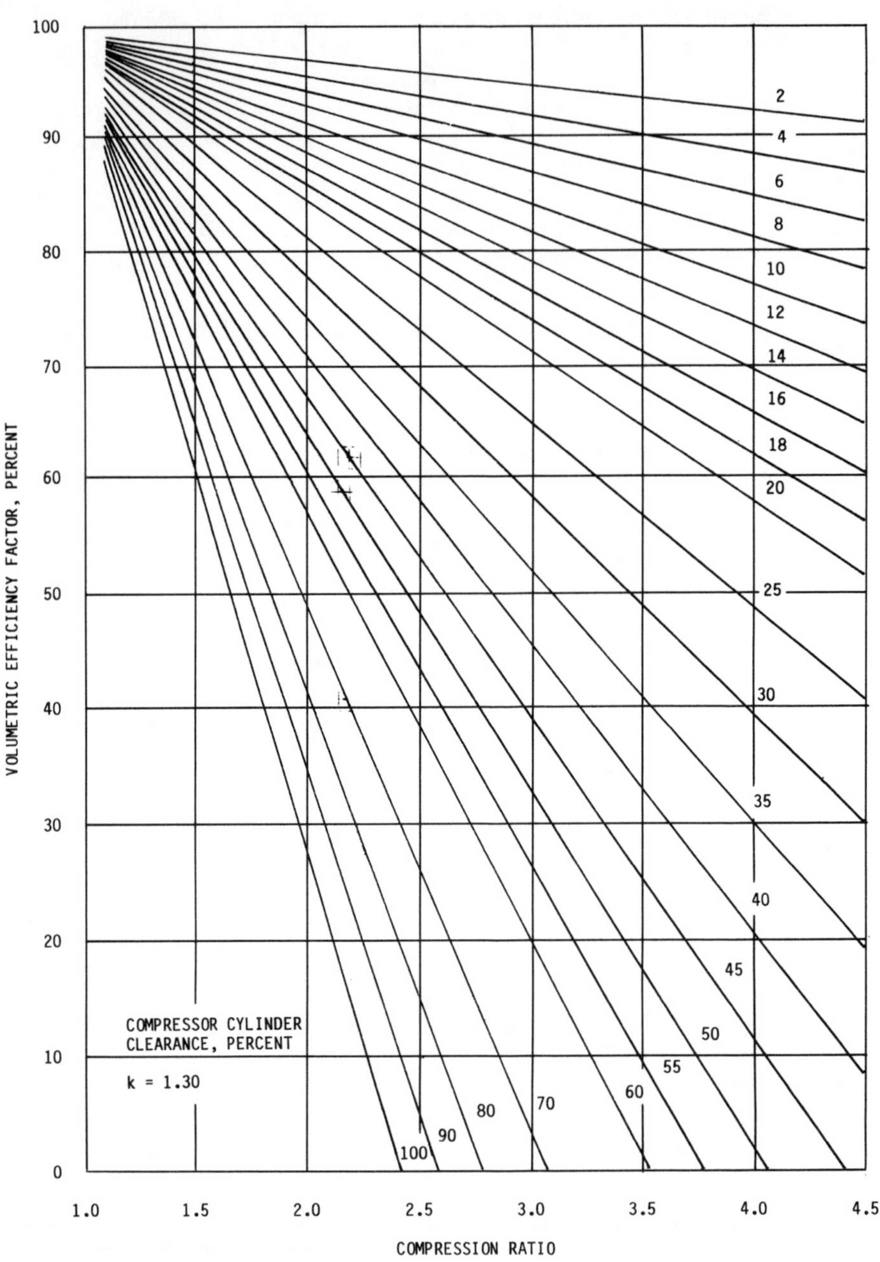


*Figure 20.* Compression ratio versus volumetric efficiency

**Compressor Cylinder Capacity.** The capacity of a specific compressor cylinder may be found from the following:

$$Q = PD \times E_V \times \frac{p_s}{p_a} \times 14.4 \times 10^{-6} \text{ MMcf/d} \qquad \text{Equation 5a}$$

or

$$Q = PD \times E_V \times \frac{p_s}{p_a} \times 864 \text{ m}^3/\text{d} \qquad \text{Equation 5b}$$

Where: $Q$ = Cylinder capacity in desired units at the prevailing inlet temperature and the base suction pressure, MMcf/d or m³/d

$PD$ = Piston displacement, ft³/min (m³/s)

$E_V$ = Volumetric efficiency factor, percent

$p_s$ = Suction pressure at the compressor flange, lbf/in² (kPa) absolute

$p_a$ = Base reference pressure, lbf/in² (kPa) absolute

If the base reference pressure is 14.4 lbf/in² (99 kPa), the cylinder capacity at this base reference pressure and prevailing temperature is:

$$Q = PD \times E_V \times p_s \times 10^{-4} \text{ MMcf/d} \qquad \text{Equation 6a}$$

or

$$Q = PD \times E_V \times p_s \times 8.73 \text{ m}^3/\text{d} \qquad \text{Equation 6b}$$

It must be emphasized that this value of $Q$ represents the actual displaced quantity of gas in MMcf/d (m³/d), corrected only for pressure to a base of 14.4 lbf/in²(99 kPa) absolute, at the prevailing suction temperature. To compare gas volumes under varying pressure and temperature conditions, it is necessary to express the volumes in standard units (MMscf) [m³/d (st)] at some pressure and temperature base. The temperature base for natural gas measurement is 60°F (15.5°C) in the United States, but there is more than one pressure base recognized by various states. Therefore, it always is advisable to check the reference conditions whenever the scf abbreviation is encountered. For standard conditions the cubic metre [m³ (st)] has only one set of reference conditions, 101.325 kPa and 15°C, as established by the international standard ISO 5024.

The recalculation of volumes from one set of reference conditions to another is rather straightforward. Reference was made earlier to pressure corrections by Boyle's Law and temperature corrections by Charles' Law. These ideal gas laws form generalized relationships between pressure, temperature, and volume of an ideal gas. In reality, gases do not behave as predicted by those laws, and it has been necessary to develop $Z$ factors, or compressibility factors, to compensate for this deviation. The compressibility factor is the ratio of the actual volume of a gas to the volume of a perfect gas predicted by the ideal gas law. If a gas occupies *less* volume at a given pressure and temperature than the volume according to the ideal gas law, the $Z$ factor is less than 1.0. Figure 21 shows the relationship between the $Z$ factor and pressure for natural gas at certain temperatures having a relative density of 0.6. The $Z$ value for gas at other temperatures or different relative density may

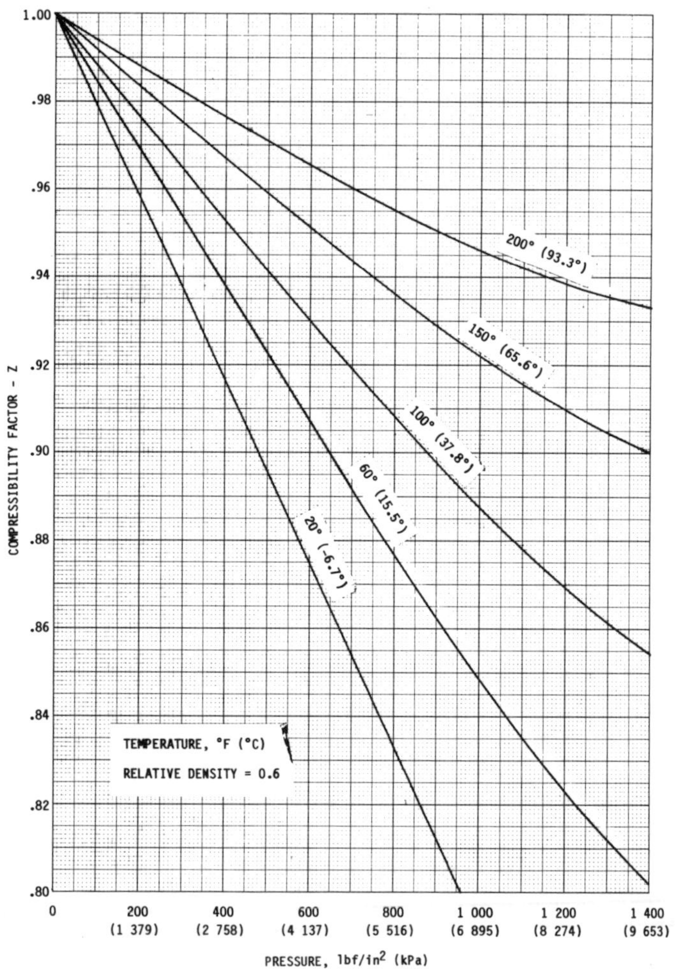


*Figure 21.* **Compressibility factor for compressor capacity calculations**

be calculated from data given in *Determination of Supercompressibility Factors for Natural Gas*, A.G.A. Catalog Number L00340. The relationship between supercompressibility factor $F_{pV}$ and the compressibility factor $Z$ is $F_{pV}=(Z)^{-0.5}$; therefore, $Z=(F_{pV})^{-2}$. The following expression also shows the volume corrections related to other pressure bases and the effect of suction temperature:

$$Q_{st} = \frac{Q \times p_a \times T_b}{Z_s \times p_b \times T_s}$$
Equation 7

Where:  $Q_{st}$ = Rate of flow, MMscf/d [m³/d (st)]
$Q$  = Rate of flow, MMcf/d (m³/d)
$p_a$ = Reference base pressure, lbf/in² (kPa) absolute
$p_b$ = Pressure base, lbf/in² (kPa) absolute
$T_b$ = Temperature base, °R (K)
$T_s$ = Suction temperature, °R (K)
$Z_s$ = Compressibility factor at inlet conditions

**Power Determination.** A number of methods are available to determine the power required to compress a given volume rate of gas between two pressure levels. The theoretical adiabatic power requirement, taking the volumetric efficiency into account, can be calculated as follows:

$$P = \frac{k}{k-1} \times \left( \frac{144 \times p_s \times PD}{33 \times 10^5} \right) \times \left[ \left( \frac{p_d}{p_s} \right)^{\frac{k-1}{k}} - 1 \right] \times E_V \; hp \qquad \text{Equation 8a}$$

or

$$P = \frac{k}{k-1} \times \left( \frac{56.85 \times p_s \times PD}{341 \times 10^3} \right) \times \left[ \left( \frac{p_d}{p_s} \right)^{\frac{k-1}{k}} - 1 \right] \times E_V \; kW \qquad \text{Equation 8b}$$

Since this result is theoretical, assuming an ideal adiabatic process, various correction factors would have to be used to compensate for pressure losses, friction, non-ideal compression and expansion, and changes in gas properties. Many reference books contain detailed formulas and methods which may be used to calculate the work and power. For the purpose of this reference, the concept of brake horsepower per million cubic feet per day [bhp/(MMcf/d)] shall be used to simplify this analysis.

Figure 22 shows a typical bhp/(MMcf/d) curve. These curves are published by various sources including compressor manufacturers, and are all similar in appearance, showing bhp/(MMcf/d) versus compression ratio for a particular $k$ value, pressure base, and temperature base.

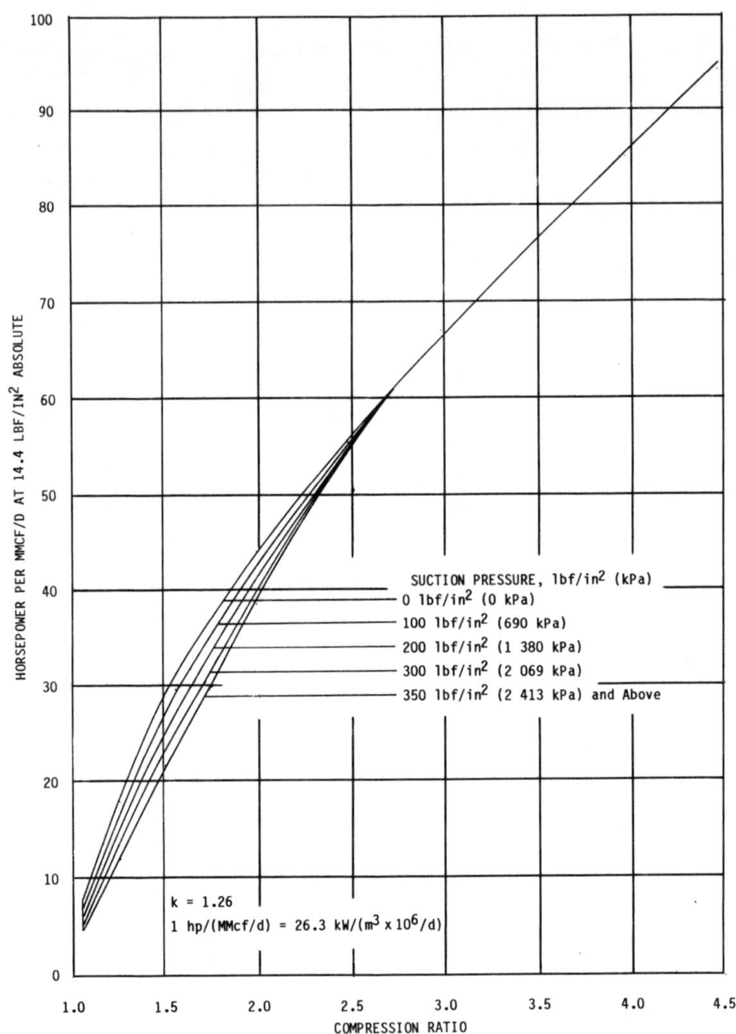


*Figure 22.* **Compression ratio versus brake horsepower per MMcf/d**

In certain instances there also are corrections to be applied to this curve that have been derived empirically to compensate for pressure losses, compression efficiencies, and other non-ideal phenomena.

Using the bhp/(MMcf/d) method, the expression for power is:

$$P = Q \times P_i \qquad\qquad \text{Equation 9}$$

Where:   $P$ = Total power required, hp
           $Q$ = Rate of flow, MMcf/d

$P_i$ = Power required per unit of volumetric flow rate, bhp/(MMcf/d)

Note that $Q$ is in MMcf/d at prevailing suction temperature and 14.4 lbf/in² absolute rather than MMscf/d. This represents the power required by the compressor cylinder to compress a given quantity of gas. The prime mover must develop slightly more power than this to overcome mechanical losses in the crossheads, connecting rods, packing cases, and piston rings. Normally the horsepower determined in Equation 9 is divided by a mechanical efficiency of 0.95 to determine the required output horsepower of an integral prime mover. For a separate prime mover, an efficiency of 0.92 frequently is used. Obviously the prime mover also must provide power for pumps, fan drives, and other parasitic loads. This mechanical efficiency is not to be confused with the compression efficiency, which is a thermodynamic term. Typical compression efficiencies for water-cooled, process-type cylinders with small gas passages is 80 to 85 percent. For air-cooled, transmission-type cylinders, typical compression efficiency is 85 to 95 percent. Figure 22 incorporates compression efficiencies for typical gas compressor cylinders. A typical modern compressor cylinder that is used in gas transmission service is illustrated in Figure 23.

## Other Factors Associated with Reciprocating Compressors

Assuming that the compressor is capable of meeting the pressure and volume conditions and that the power requirement is within the physical capability of the prime mover and the compressor frame, it is necessary to consider five other factors.

**Maximum Working Pressure.** The maximum working pressure (MWP) is the highest discharge pressure which can be developed within the cylinder without exceeding the manufacturer's design limitations. Each manufacturer is specific as to the MWP rating and relief valve setting, which usually are stamped on a nameplate. Usually compressor cylinders are made of cast iron, nodular iron, or forged steel and have different allowable working pressures, depending upon the class (size) and material. Typical working pressure ranges are as follows:
- Cast iron cylinders, large diameter [20–36 in. (508–914 mm)] <200 lbf/in² (1 380 kPa)
- Nodular iron cylinders, medium diameter [10–20 in. (254–508 mm)] <1 300 lbf/in² (8 965 kPa)
- Forged steel cylinders, small diameter >1 300 lbf/in² (8 965 kPa)

**Rod Load.** Rod load, a term which describes another important consideration in compressor sizing, is directly proportional to the pressure

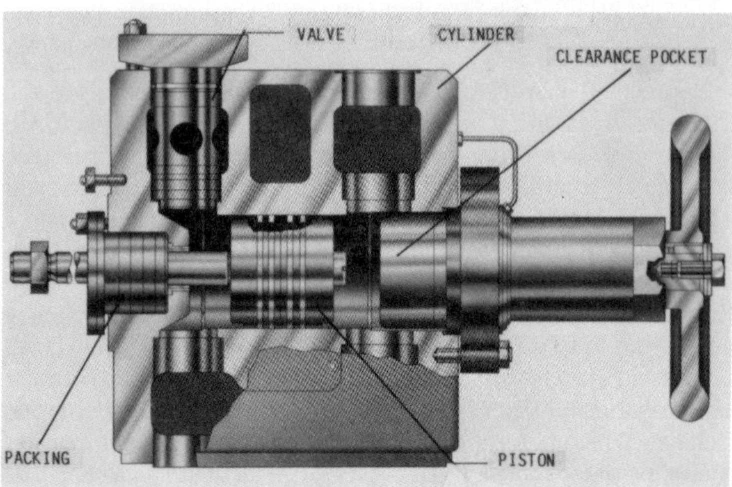


**Figure 23.** **Cross-section of a modern pipeline cylinder**

differential ($p_d$–$p_s$). High pressure on one side of the piston face and a low pressure on the other side subjects the piston rod to either a tensile or compressive stress, depending upon which end of the piston the highest pressure is acting. The force acting on the rod is calculated from the respective pressures, cylinder bore, and rod area. The rod area is necessary because of the resulting reduction in the piston face area on the crank end.

Many times the manufacturer's specified rod load limit is not related necessarily to the strength of the rod, but to some other part of the machine such as the crosshead, connecting rod, or crankshaft. Normally the tension and the compression limits are specified. Actual rod loads may be calculated from the following:

$$F_T = A_H \times (p_d - p_s) - A_R \times (p_d - p_{amb}) \qquad \text{Equation 10}$$

and

$$F_C = A_H \times (p_d - p_s) + A_R \times (p_s - p_{amb}) \qquad \text{Equation 11}$$

Where:   $F_T$   = Tension force in desired units, lbf or N
$F_C$   = Compression force in desired units, lbf or N
$p_d$   = Discharge pressure, lbf/in² (Pa), absolute
$p_s$   = Suction pressure, lbf/in² (Pa), absolute
$p_{amb}$ = Atmospheric pressure, lbf/in² (Pa), absolute
$A_H$   = Area of the head end of the piston, in² (m²)
$A_R$   = Cross-sectional area of the rod, in² (m²)

**Discharge Temperature.** Gas temperatures increase as gas is compressed. In a reciprocating compressor where velocities, turbulence, and slip are low (all of which cause entropy gains), the following equation is used to calculate the discharge temperature:

$$T_{\mathrm{d}} = T_{\mathrm{s}} \times \left( \frac{p_{\mathrm{d}}}{p_{\mathrm{s}}} \right)^{\frac{k-1}{k}} \qquad\qquad \text{Equation 12}$$

Where: $T_{\mathrm{d}}$ = Absolute discharge temperature, °R or K
$T_{\mathrm{s}}$ = Absolute suction temperature, °R or K

While the discharge temperature of most water-cooled reciprocating compressors operating at a volumetric efficiency greater than approximately 20 percent is close to that calculated by the above isentropic equation, it may differ considerably for others. In particular, the discharge temperature with non-cooled cylinders will be higher because of the higher ratio imposed by the valve pressure drop and by irreversibility (entropy gain). Compressors with low piston speeds generally exhibit a greater cylinder cooling effect than do high-speed machines since the longer gas residence time in the cylinder permits higher heat transfer. Compressors operated at pressure ratios above 3.5 generally exhibit lower $T_{\mathrm{d}}$ than calculated because the higher discharge temperature results in a higher gradient with the ambient, whereas ratios less than 2.5 result in a $T_{\mathrm{d}}$ higher than isentropic.

To more accurately predict the discharge temperature, it may be advisable to modify the $(k-1)/k$ factor by dividing it by the polytropic efficiency. This efficiency is derived empirically from the actual compressor process (primarily from shop or field tests), and a new term — polytropic exponent $n$ — is used to replace $k$ in the above formula.

Discharge temperatures are controlled primarily by limiting the compression ratio across the compressor cylinder. In reality it is desirable to maintain discharge temperatures below 250°F (121°C), although there are some valve and ring materials which can be used continuously between 250 and 300°F (121 and 149°C). Air-cooled, pipeline-type compressor cylinders have maximum discharge temperature limitations in a range of 150 to 200°F (66 to 93°C). Water-cooled cylinders are used at high discharge temperatures to keep the cylinder temperature uniform, thereby maintaining dimensional stability. Water cooling is not intended to reduce the heat of compression, although it does help to ensure adequate viscosity in the lubricant flow.

**Piston Speed.** Piston speed is a measure of the average rate that the piston and rod travel within the cylinder. The wear rate of these components is related directly to the piston speed, which is determined by:

$$v = \frac{N \times l}{6} \text{ ft/min} \qquad\qquad \text{Equation 13}$$

or

$$v = \frac{N \times l}{30} \times 10^{-3} \text{ m/s}$$

Where:   $v$ = Piston speed in desired units, ft/min or m/s
          $l$ = Length of stroke, inches (millimetres)

Piston speeds typically range from 800 to 1 100 ft/min (4.1 to 5.6 m/s).

**Rotational Speed.** Rotational speed affects both the piston speed and the piston displacement. It is also a factor in valve life because of the frequency of the valve opening/closing cycles. Compressor speed may be classified as low (below 400 r/min), medium (400–800 r/min), or high (above 800 r/min). Rotational speed may be varied within the limits of the prime mover in order to vary the quantity of gas compressed.

Compressor operating characteristics incorporating the preceding concepts will be discussed under the heading Compressor Characteristics and Load/Capacity Control.


## Dynamic Compressor Calculations

The most commonly used dynamic machine in the natural gas industry is the centrifugal compressor, such as the one shown in Figure 24. This machine consists of one or more radial flow impellers mounted on a shaft which rotates within a casing, as illustrated in Figure 25. Stationary parts, such as inlet guide vanes, diffuser passages, turning vanes, and outlet guide vanes, direct the flow of gas through the compressor. Each group of vanes, impeller, and diffuser constitute a stage of the compressor. The gas is compressed to the required pressure level by receiving energy as it passes through the stages of compression, and is directed to the eye (or center) of the impeller by the inlet guide vanes. Energy is imparted by the impeller to the gas which leaves the periphery of the impeller at a high velocity and enters either a vaned or vaneless diffuser. In the diffuser the high kinetic energy of the gas is converted into additional static pressure, or potential energy. Since there are pressure losses within the compressor components, the impeller must impart sufficient energy to satisfy the pressure requirements and overcome these losses. These losses determine the compression efficiency which is an important factor affecting compressor operation and economics.

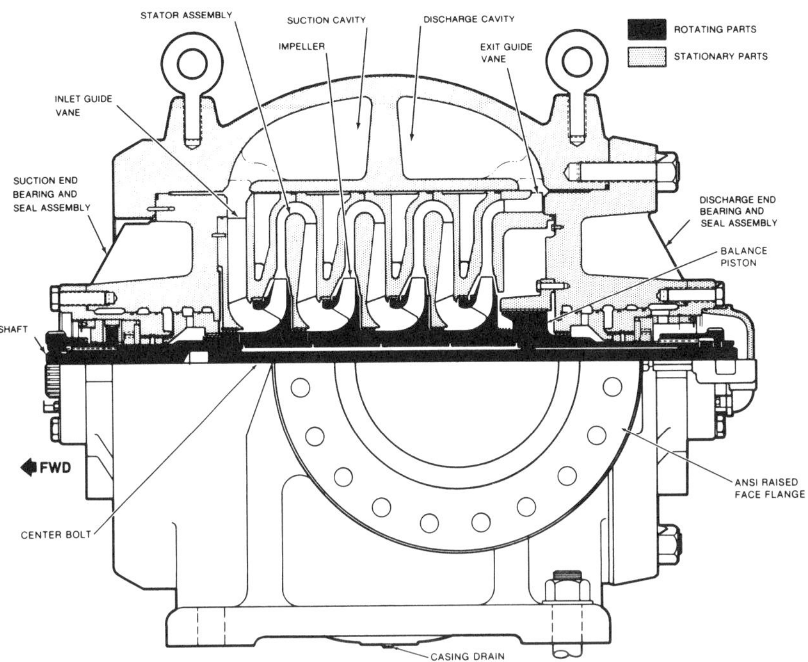


**Figure 24.** Cutaway of centrifugal compressor

**Figure 25.** Cutaway view of centrifugal compressor

## Dynamic Compressor Performance

The performance of a centrifugal compressor usually is measured and described in terms of head, volume flow, efficiency, and power.

**Head.** Head is a measure of the energy input per unit mass into the gas stream by the compressor. It is produced by the velocity changes in the gas flow resulting from the action of the impeller. Since the compressor geometry remains constant (as contrasted with a reciprocating or positive-displacement compressor), a variation of flow will cause changes in velocity. This in turn causes a varying head. A compressor operating on a given gas at a fixed speed will produce a head versus volume flow relationship that essentially remains constant; that is, for each value of flow there is a corresponding value of head. Since head is produced by velocity changes that are dependent upon volume rather than gas properties or conditions, the characteristic head versus volume curve is independent of the gas involved. Although the head versus volume characteristic can be considered constant, the pressure ratio and volume changes vary greatly from one gas to another.

Theoretical head may be calculated as follows:

$$H_a = Z_s \times R_c \times T_s \times \frac{k_m}{k_m - 1} \times \left[ \left( \frac{p_d}{p_s} \right)^{\frac{k_m - 1}{k_m}} - 1 \right] \qquad \text{Equation 14}$$

Where: $H_a$ = Adiabatic head in desired units, ft•lbf/lb or J/kg
$\quad\quad\quad$ $Z_s$ = Compressibility factor at compressor suction conditions
$\quad\quad\quad$ $k_m$ = Mean ratio of specific heat capacities, $c_p/c_V$
$\quad\quad\quad$ $R_C$ = Specific gas constant for a particular gas
$\quad\quad\quad\quad$ 1 545.3/$M$ ft•lbf/(lb•°R), [8.314 41/$M$ J/(kg•K)], and
$\quad\quad\quad$ $M$ = Molar mass of gas, lb/(lb•mol), (kg/mol)

The mean ratio of specific heat capacities is the average $k$ value that occurs throughout the compression process. Since the specific heat capacities, $c_p$ and $c_V$, are changing constantly as the temperature of the gas is increasing, the $k$ value at discharge conditions can be substantially different than at suction conditions. One method of finding the mean ratio of specific heat capacities, $k_m$, is to assume a discharge temperature and calculate $k$ at the discharge conditions. Then $k_m$ can be found by calculating the numerical average of $k$ at suction and discharge conditions.

The resulting value of $k_m$ then should be used to calculate the theoretical discharge temperature, followed by calculating a new value of $k$ at discharge conditions. A second value of $k_m$ then is calculated and the entire process is repeated. This iteration continues until the margin of error is within desired limits, and the last value of $k_m$ is used in Equation 14. (This method also may be used in Equations 4 and 8 for reciprocating compressor calculations.)

This calculation yields the adiabatic head, which assumes that no heat was transferred into or out of the gas during the compression process. As with the reciprocating compressor, the value of the polytropic exponent $n$ may be substituted for $k$. This will yield a polytropic head, which more closely will represent the actual process. The polytropic exponent must be developed experimentally. It can be seen from the above equation that the head is dependent upon the ratio of specific heat capacities, molar mass of the gas, inlet temperature, and compression ratio. Conversely, the molar mass of the gas, inlet temperature, compressibility factor, and $k$ value determine the pressure ratio resulting from the head developed by a stage of compression.

**Volume Flow Rate.** As indicated above, variations in volume flow rate affect the head developed by an impeller. The specific shape of the head-flow curve is a function of the stage geometry, that is, the inlet guide vane angle, impeller blade angle, and outlet diffuser construction. When describing centrifugal compressor performance it is necessary to convert the mass flow in lb/h (kg/h), or volume flow in MMscf/d [m³/d (st)], to actual volumetric rate at the inlet conditions. This is the volume of gas that actually is entering the eye of the impeller.

To convert the volumetric flow rate from reference conditions to suction conditions, the following is used:

$$q_i = Q_{st} \times \frac{p_b}{p_s} \times \frac{T_s}{T_b} \times Z_s \times \frac{10^3}{1.44} \ \text{ft}^3/\text{min} \qquad \text{Equation 15a}$$

or

$$q_i = Q_{st} \times \frac{p_b}{p_s} \times \frac{T_s}{T_b} \times Z_s \times \frac{10^{-3}}{86.4} \ \text{m}^3/\text{s} \qquad \text{Equation 15b}$$

Where: $q_i$ = Actual inlet volumetric flow rate in desired units, ft³/min or m³/s

$Q_{st}$ = Volumetric flow rate at reference conditions, MMscf/d [m³/d (st)]

When the original data is reported as mass flow per hour, the actual volumetric flow rate can be determined by multiplying the mass flow rate by the specific volume value [ft³/lb (m³/kg)] and dividing it by the appropriate factor to account for differences in time base [60 min/h (3 600 s/h)].

**Efficiency.** As with any machine the ideal compressor performance is affected by a number of factors. There are internal friction losses which are proportional to the square of the flow. There are also incidence losses due to the angle of attack of the gas stream at the impeller inlet. If this angle does not coincide with the blade entrance angle, there is a velocity component which is wasted, and appears as head loss. Other losses, such as boundary layer separation and wakes and slippage, affect the characteristic curve as well as changes in efficiencies that vary with the head, flow, and speed. The basic shape of the curve is a function of the various angles and geometry of the fixed and moving components of the compressor. Manufacturers of centrifugal compressors usually express efficiency in one of two ways, adiabatic or polytropic. The performance analysis of a given compressor is not affected by a particular efficiency as long as polytropic head is used with polytropic efficiency, and adiabatic head is used with adiabatic efficiency. Because of turbulence associated with high velocities employed in centrifugal compressors, the irreversibility effect (entropy gain) is high. Therefore, polytropic theory is particularly applicable to these machines. The determination of compressor efficiencies can be made only by accurate test methods utilizing pressure, temperature, and flow data.

**Power.** The theoretical equation for gas horsepower (kilowatts) for a centrifugal compressor is:

$$P = \frac{144 \times d \times p_s \times q_i \times H_a}{53.35 \times Z_s \times T_s \times 33\ 000} \text{ hp} \qquad\qquad \text{Equation 16a}$$

or

$$P = \frac{d \times p_s \times q_i \times H_a}{287 \times Z_s \times T_s} \text{ kW} \qquad\qquad \text{Equation 16b}$$

Where:  $P$ = Input power to the gas stream in desired units, hp or kW

And:    $d$ = Gas relative density

To account for the aerodynamic losses, it is necessary to replace $H_a$ with $H'/\eta'$, where $H'/\eta'$ is the adiabatic or polytropic head divided by the corresponding adiabatic or polytropic efficiency.

In addition to the aerodynamic losses, there are mechanical losses due to seal and bearing friction. For service design purposes, the mechanical losses for centrifugal and rotary compressors over 1 000 hp (746 kW) can be assumed to be 35 hp (26 kW) for bearings and 35 hp (26 kW) for oil-type shaft seals; below that size, losses will amount to 1 to 3 percent. When possible, the vendor should be consulted. In any case, these losses are included in the mechanical efficiency factor. Therefore, the required shaft power is:

$$P_s = \frac{P}{\eta_m}$$

Equation 17

Where: $P_s$ = Compressor shaft input power in desired units, hp or kW

$\eta_m$ = Mechanical efficiency (0.97 to 0.99)

Since centrifugal compressors typically rotate at high speeds (5 000 to 25 000 r/min), a gearbox may be required when the prime mover is an electric motor or a reciprocating gas engine. If there is a gearbox between the prime mover and the compressor, the transmission losses, as well as the input power requirements for any direct-driven accessories, must be considered in determining the required prime mover power output.

## Other Factors Associated with Dynamic Compressors

In addition to the compressor attaining the desired pressure and volume within the power producing capabilities of the prime mover, there are three other factors which must be considered in the discussion of dynamic rotating compressors. They include stability, maximum working pressure, and discharge temperature.

**Stability.** For each distinct value of rotational speed there is a relationship between the volume flow through the compressor and the head, produced by a stage of compression. Also, there is a minimum volume flow for each speed—below which a condition called surge occurs. Surge, a highly unstable condition, results in the entire flow momentarily reversing from discharge to inlet. This reduces the discharge pressure, and the compressor again is able to pump into the system. This causes the discharge pressure to increase again, and the cycle repeats. Such oscillation causes severe machine and piping vibration, and may result in damage.

A condition opposite to surge is choke, or stonewall. Choke is the maximum flow that a compressor can handle at a given speed. The high flows cause large internal losses, and the compressor is unable to produce any net pressure increase. All the energy supplied to the compressor is dissipated in losses. Unlike surge, choke will not damage the compressor.

**Maximum Working Pressure.** Maximum working pressure (MWP)

is the highest discharge pressure which can be developed within the compressor casing without exceeding the manufacturer's design. Rotating compressor casings may be either cast or fabricated. Fabricated casings are essentially cylinders which can be designed for extremely high pressures. Pressure ratings of dynamic compressors used in the gas industry are generally in the 1 000 to 1 500 lbf/in² (6 900 to 10 350 kPa) range. Higher pressure applications and/or high compression ratios usually require multistage process compressors, which are typically smaller and less efficient.

**Discharge Temperature.** The relationship between the discharge temperature and the compression ratio is the same for dynamic compressors as for reciprocating compressors as described above. Since dynamic compressors generally have a lower thermodynamic efficiency than reciprocating compressors, the polytropic exponent $n$ should be used in place of $k$. For gas with a $k$ value of 1.30, the centrifugal compressor process may yield an $n$ value of 1.38. This will result in a slightly higher discharge temperature than that calculated for a reciprocating compressor for any given compression ratio.

## COMPRESSOR CHARACTERISTICS AND LOAD/CAPACITY CONTROL

This discussion describes the operational characteristics of the reciprocating and dynamic compressors used in the natural gas industry. Methods of load and capacity control also are discussed.

### Reciprocating Compressor Capacity Control

Reciprocating compressors are used widely in the natural gas industry for all types of services and applications. They are available in sizes up to 15 000 hp (11 190 kW) and can operate efficiently under a wide range of volume and pressure conditions.

Reciprocating compressor limitations have been described previously in Other Factors Associated With Reciprocating Compressors. The limitations of maximum working pressure, rod load, and discharge temperature may be controlled by limiting the compression ratio.

If the required compression ratio for the pipeline conditions exceeds that attainable by a single unit with its cylinders operating parallel, it will be necessary to operate the compressors on the unit in series, or stages, so that each compressor will be operating within its limit. The unit is said to be two-staged if the total compression ratio is achieved in two steps. The number of stages is limited only by the number of cylinders that can be mounted on the compressor frame. Staging requires that the discharge gas from the first stage be cooled before it enters the suction of the second or subsequent stages. Without intercooling, the gas temperatures would be compounded through the various stages.

From a thermodynamic standpoint, multiple compression stages with intercooling are slightly more efficient than a single stage of compression, but pressure drops in the intercoolers and connecting piping and the higher cost of cooling and piping equipment have an adverse impact on the economics of staging. Unless provisions are made in the original design, multistaging usually requires modifications to gas piping and compressor cylinders.

Piston speed and rotational speed also must be considered when using reciprocating compressors. From the basic equations it can be seen that the capacity of a compressor is related directly to piston displacement, which in turn is related to speed; that is, for the same cylinder size, a high-speed compressor unit delivers greater volumes than low-speed units deliver. Also, for a given power output, a high-speed prime mover is physically smaller than a low-speed prime mover. Generally, high-speed compression equipment is smaller and less expensive to purchase and install than low-speed equipment.

High-speed compressors, however, may have lower compression efficiencies due to pressure losses in the valves and gas passages of the small cylinders. Also, the wear rate of components operating at high rotational and linear speeds may be higher than those of similar components of low-speed machines. It also should be noted that compressor rod load can be affected indirectly by speed; that is, the cylinder area may be optimized by design to reduce rod load without sacrificing piston displacement if the speed can be increased.

### Compressor Cylinder Clearance

As previously discussed, the power required to deliver a given volume rate of gas is directly proportional to the compression ratio and the quantity of gas compressed. Unlike an air compressor, which generally operates between a fixed suction and slightly varying discharge pressures, a gas compressor operates with variations in both suction and discharge pressures. Depending upon the type of service, pressure variations may be quite large. If the pressure variations cause variations in the compression ratio, the bhp/(MMcf/d) value will change. To limit the power requirement when the compression ratio increases, it is necessary to reduce the volumetric capacity of the cylinder. Equation 5 states that cylinder capacity will be reduced if volumetric efficiency, piston displacement, and/or suction pressure are reduced.

A commonly employed method is to change the volumetric efficiency by adding or removing clearance volume. This is done in a number of ways, but the most common is through the use of clearance pockets such as those shown in Figure 26. These pockets are castings of specific volumetric capacity mounted on the compressor valves or in the cylinder heads. A passage through the center of the valve with a movable plug at the clearance pocket end permits the clearance pocket volume to be added to the fixed compressor clearance. This clearance is opened or closed as dictated by the operating pressures. Not only does this control prevent overload, but it also prevents an underload condition when the suction and discharge pressures vary. This method of load, or torque control, regulates the prime mover power requirements to the most fuel efficient level.

It should be emphasized that the addition or removal of clearance to control compressor torque has an effect on one of two components of power. The other component, which is speed, can be varied also—which in turn varies the piston displacement of the compressors. While a reduction in speed will result in a reduction in power, it does not result necessarily in a reduction in torque. A variation in speed without a variation in compressor clearance generally causes the suction and discharge pressures to change, which, in turn, will cause either over-torque or under-torque.

Another method of controlling cylinder delivery is to reduce piston displacement by deactivating one end of a double-acting compressor. Generally this is accomplished by removing a suction or discharge valve or by operating a valve lifter, or unloader, to restrain the action of the valve plates and prevent gas from being trapped in the cylinder.

### Reciprocating Compressor Operating Curves

The characteristics of a particular compressor may be represented by an operating curve. Various formats may be used, but all serve to provide information to the operator on capacity and power levels for various clearance and pressure conditions. Figure 27 shows a typical operating curve using a thumbprint format. The numbered regions on the curve are called loading steps and represent amounts of clearance volume which must be in effect to control the compressor torque. For each suction and discharge pressure, there is a specified amount of clearance required. Once the proper compressor step is determined, Figure 28 is used to determine the amount of gas being compressed and the horsepower developed at specific pressure conditions. There is a step curve for each clearance condition. If the compressor is operating at less than rated speed, the volume and horsepower are multiplied by the ratio of actual speed to rated speed. The reference curves are marked to illustrate operation at a suction pressure of 290 lbf/in² (2 000 kPa) gauge and a discharge pressure of 510 lbf/in² (3 520 kPa) gauge.

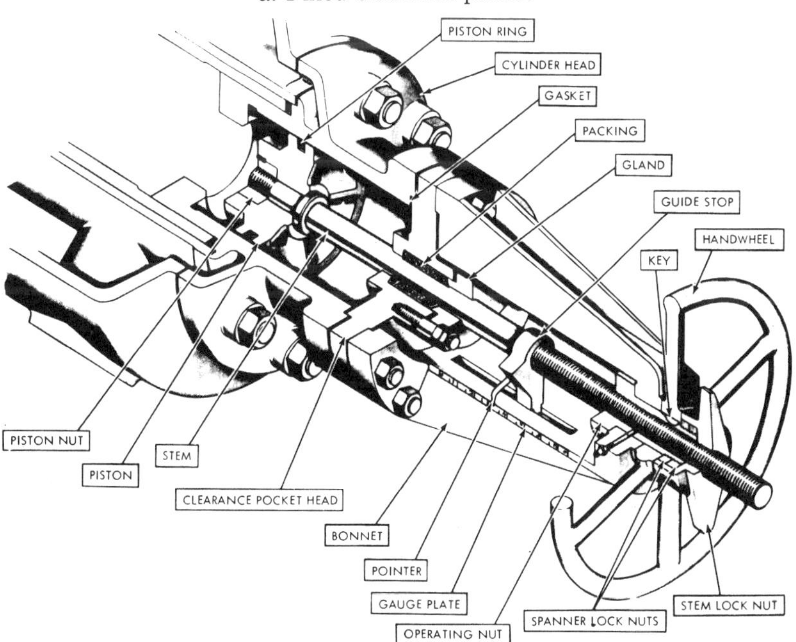


a. Fixed clearance pocket

b. Variable clearance pocket

***Figure 26.*** **Compressor cylinder head end clearance pocket configuration**

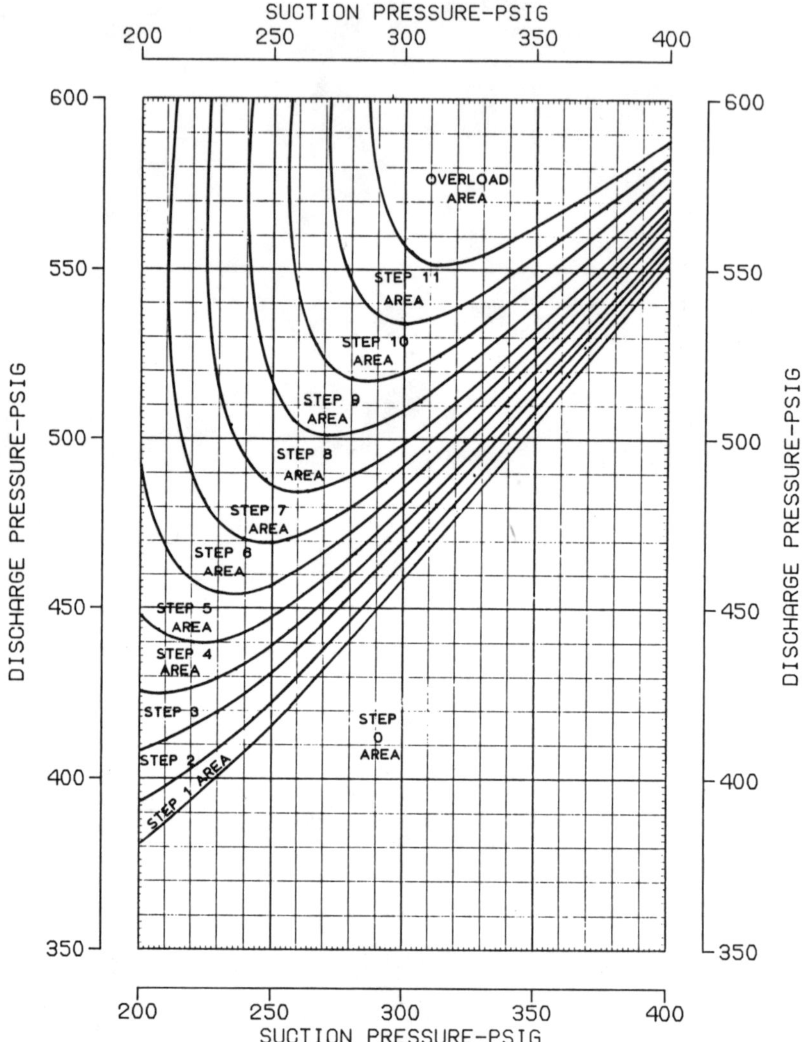


*Figure 27.* **Typical reciprocating compressor cylinder loading diagram**

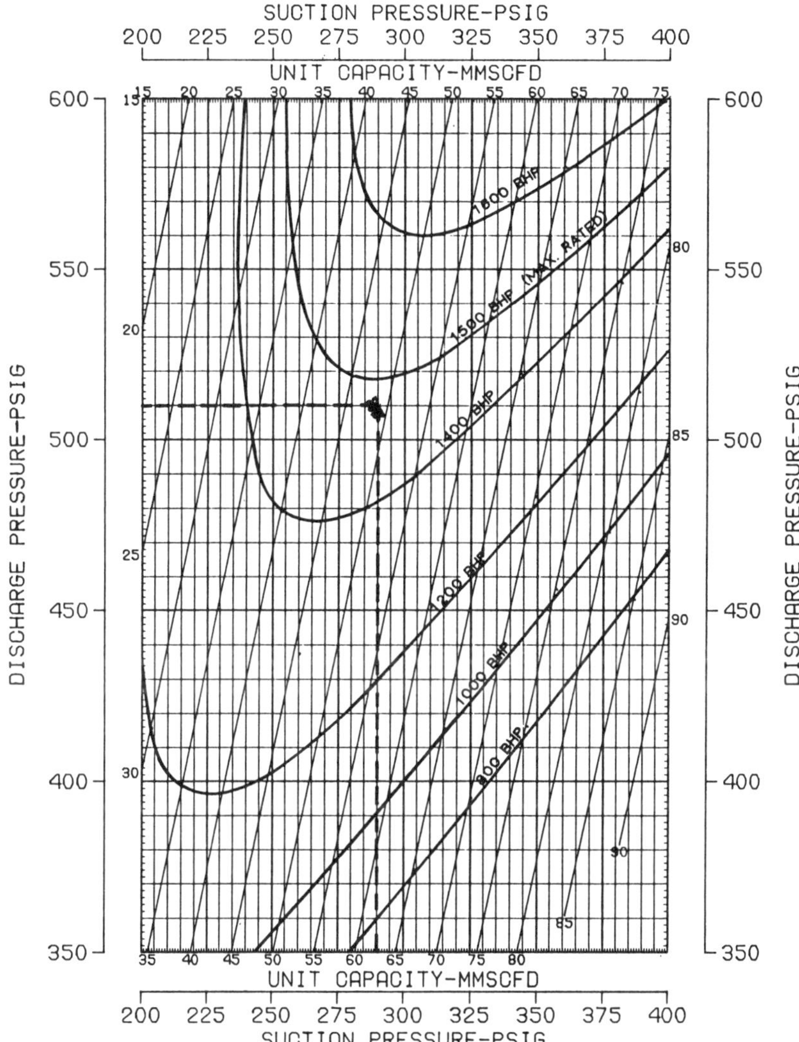


*Figure 28.* Typical horsepower-capacity diagram for reciprocating compressor

## Dynamic Compressor Capacity Control

Dynamic compressors do not have the wide range of applicability in the natural gas industry that reciprocating compressors have. However, there are several factors that permit centrifugal compressors to be used successfully in high-volume, low-compression-ratio service. As discussed earlier under Other Factors Associated With Stability of Rotating Compressors, there are stability limitations on rotating compressors that are not inherent in reciprocating compressors. On the other hand, centrifugal compressors driven by gas turbine prime movers in the 1 000- to 30 000-hp (746- to 22 380-kW) range can handle large volumes of gas with a low initial capital investment, primarily due to the compact construction of the turbine/compressor package. Lower compression and prime mover efficiencies require that capital cost savings be evaluated carefully, particularly for a high-load-factor operation.

Operating limitations for centrifugal compressors are somewhat analogous to those for reciprocating compressors except that the compression ratio aspect becomes more critical. A centrifugal compressor contains one or more impellers, or stages. Large, high-flow units usually contain one or two stages, and can operate efficiently with compression ratios in the 1.1 to 2.0 range. Higher compression ratio requirements are met either by compressors connected in series or by using smaller, multistage units. The latter alternative is slightly less efficient, with the overall efficiency being inversely proportional to the number of stages. The lower compressor efficiency results in discharge temperatures being higher than those calculated for reciprocating compression.

Since one of the major concerns in centrifugal compressor operation is surge, a system has been devised to protect the compressor unit in the event the volume flow is reduced below the allowable level. The surge-control system compares the actual flowing volume with that required for a particular head. If the actual volume is less than that required to prevent surge, a bypass valve opens to permit a portion of the discharge gas to flow into the suction, thereby increasing the volume flow through the compressor. This is inefficient, but necessary to protect the machine.

Flow rate can be varied through a centrifugal compressor by regulating compressor speed. This variation is dependent on compressor geometry since reduced speed reduces the compressor's ability to develop head. Another performance constraint is the ability of the prime mover to develop sufficient power at the speed required to prevent the unit from surging.

### Centrifugal Compressor Operating Curves

The characteristics of a centrifugal compressor are represented also by an operating curve. Unlike the reciprocating compressor, the ability of a dynamic compressor to develop the required head is dependent on flow. The interdependence of these two variables complicates the operating curve. Figure 29 illustrates this with a typical curve marked to indicate the operation with a suction pressure of 600 lbf/in² (4 140 kPa) gauge and a discharge pressure of 802 lbf/in² (5 530 kPa) gauge. Also associated with the volume required to produce the head is an operating speed, and the resultant power input requirements. For a variable speed prime mover such as a gas turbine, it is important to make certain that the available power (at the prevailing ambient temperature) is sufficient for the compressor requirements.

The proper application of the compressor equipment to the required service is important, as it results in the economical and reliable operation of the gas transmission system. When all the appropriate factors have been considered, the compressor will be able to operate over variations in pipeline pressures and flows, and will result in minimized fuel usage. Also, optimum installation, operation, and maintenance costs will be achieved.

# ENGINE AND COMPRESSOR AUXILIARY EQUIPMENT

This topic covers the basic fundamentals involved in the design of the major auxiliary equipment used in a natural gas compressor station. Because of the many types of engines/compressors and the various environments where these units are installed, only the more typical design considerations will be covered.

## AIR INLET SYSTEM

Air to be used for the combustion of fuel in an internal combustion engine or a gas turbine must be provided in the right quantity and quality. The air inlet system primarily consists of an inlet air filter and the piping, or ducting, to direct the air to the engine. Silencers also may be required to attenuate the noise created by the flow of the air and by the engine itself. Ambient rated, turbocharged engines, as well as gas turbines, also require some means of controlling the temperature of the air entering the engine if full power is to be developed.

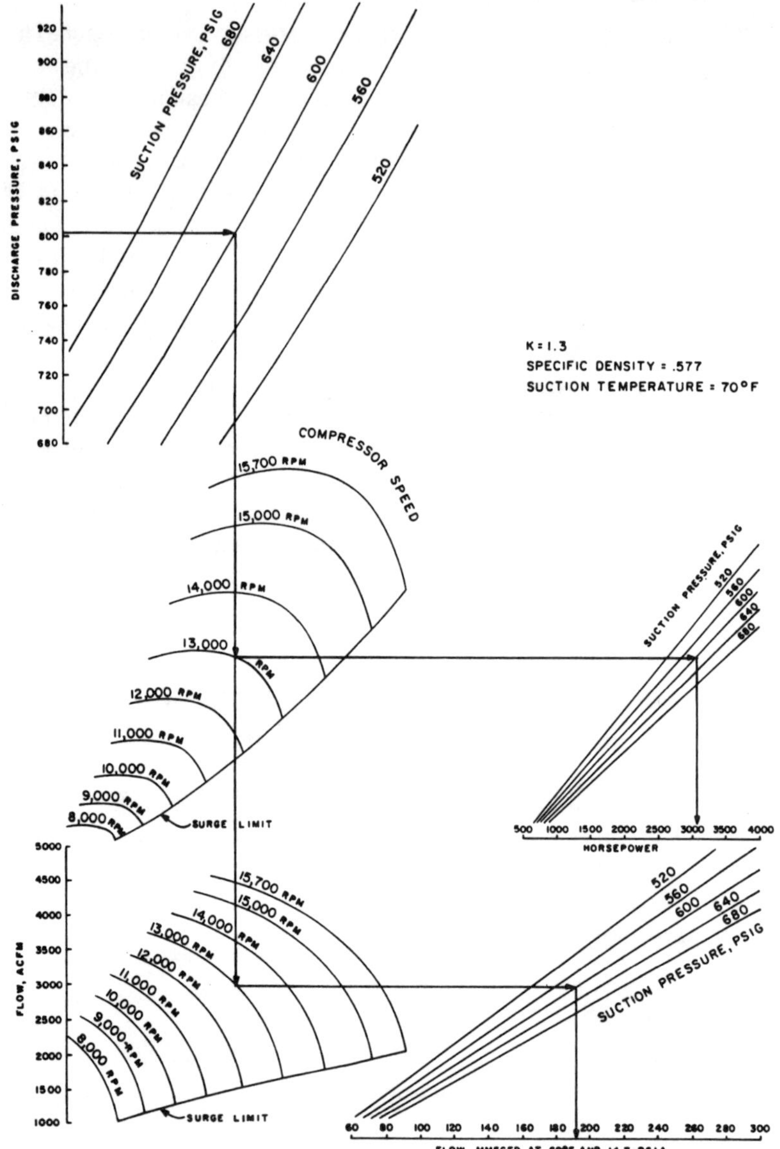


*Figure 29.* **Typical centrifugal compressor flow nomograph**

## Air Filtration

Some situations and environmental conditions require installation of a wet type filter. Because of the ease in maintenance (changing filter elements) and good filtering with minimum pressure drop, however, the dry type is being used everywhere else.

Figure 30 shows a dry type filter which is basically a housing containing the dry filter elements. These elements are generally of pleated paper or molded fiber glass; and they are the fixed tubular, panel, or movable roll type. The movable roll type air filter is used primarily with engines or turbine units requiring very high air flows. Wet type filters generally utilize a wire mesh or screen material as a filter element through which the air passes after coming in contact with a bath of fluid, usually light oil. When heavy particulates are present, inertial separators may be used on gas turbines.

*Courtesy of United Gas Pipe Line Company*

***Figure 30.*** **Dry type air filter**

When selecting an air filter, the ambient air conditions regarding dust particle size and chemical contaminants should be considered as well as ambient temperature and weather conditions. To select the proper filter housing, the following must be considered: type of filter, volume of air to be handled, protection against corrosion, wind and snow loads, accessibility for maintenance, the need for cooling and silencing, and the pressure drop through the filter.

## Inlet Air Piping

Inlet air piping from the air filter to the engine must be sized so that the pressure drop does not exceed the engine manufacturer's allowable limit. The pipe size should be at least as large as the inlet air connection to the engine. The most efficient and economical system is one with the minimum distance between the filter and the engine, thus minimizing the pipe size, the number of bends used in the piping system, and pressure loses.

Once the piping layout has been determined, the theoretical pressure drop in the system can be calculated.

## Inlet Air Silencing

In most intake air systems for internal combustion engines, the inlet filter can be supplied with sound attenuating qualities. The maximum allowable noise level at some specified distance from the source should be specified to the air filter manufacturer.

Inlet air systems for large reciprocating engines and gas turbines handle high velocities resulting in noise levels that may not be attenuated adequately by the air filter. In these cases the installation of an inline silencer downstream of the air filter is an effective method of reducing the noise to an acceptable level. Pipe or ducts also may be lagged, or encased, with sound insulation to attenuate the noise radiating from the pipe walls, but this generally is not very effective at the lower end of the audible range. The pressure drop across the silencer also must be taken into consideration when calculating the total pressure drop of the intake air system.

## Inlet Air Temperature Control

For turbocharged engines it is necessary to control the air temperature entering the engine in order to maintain optimum efficiency over the range of varying ambient air temperatures. This is accomplished by installing a preheater upstream of the turbocharger and/or an aftercooler downstream of the turbocharger. By means of a temperature regulator valve, the preheater maintains the temperature of the inlet air above a minimum, which is generally 40°F (4.5°C). Because of the high ambient temperatures and the temperature rise across the turbocharger, it is necessary to cool the air before it enters the engine so that the air manifold temperature is within the requirements set by the engine manufacturer. A typical engine air manifold temperature range would be 70 to 120°F (21 to 49°C). This relates to an ambient air temperature of 40 to 100°F (4 to 38°C). The control of this temperature operates in conjunction with the air/fuel ratio control to provide an efficient operating engine through the varying load conditions that may be experienced.

## EXHAUST SYSTEM

The exhaust system of an engine or turbine provides a passageway for the exhaust gases to pass from the engine to the atmosphere. This passageway must be designed to prevent excessive back pressure on the engine and to attenuate the noise created by the exhaust gas velocities and pulsation effect of a reciprocating engine. Also, the heat radiation due to the high temperature exhaust gases must be controlled to prevent adverse effects to personnel and the surroundings. This is done through proper thermal insulation of the exhaust piping.

### Exhaust Heat Recovery

The exhaust gases leaving the engine or turbine range in temperatures from 500°F (260°C) to over 1 000°F (540°C). This hot gas represents a large portion of the energy from the fuel gas burned in the engine that did not perform any useful work, and at most installations is lost to the atmosphere. However, it can be recovered and utilized to do useful work.

To recover the heat, the exhaust gas can be piped to a waste heat boiler where the heat produces steam or hot water, or may be used to heat another transfer medium. The steam or heat fluid can be used in many processes and plants where the energy is needed. Some typical uses of recovered heat are:
- Electric power generation
- Process steam in chemical plants and refineries
- Controlled environment hot houses for growing food crops
- Steam turbine drives for combined-cycle compressor units
- Steam injection into a gas turbine to increase the power output
- Manufacturing facilities
- Glycol regeneration, as shown in Figure 31.

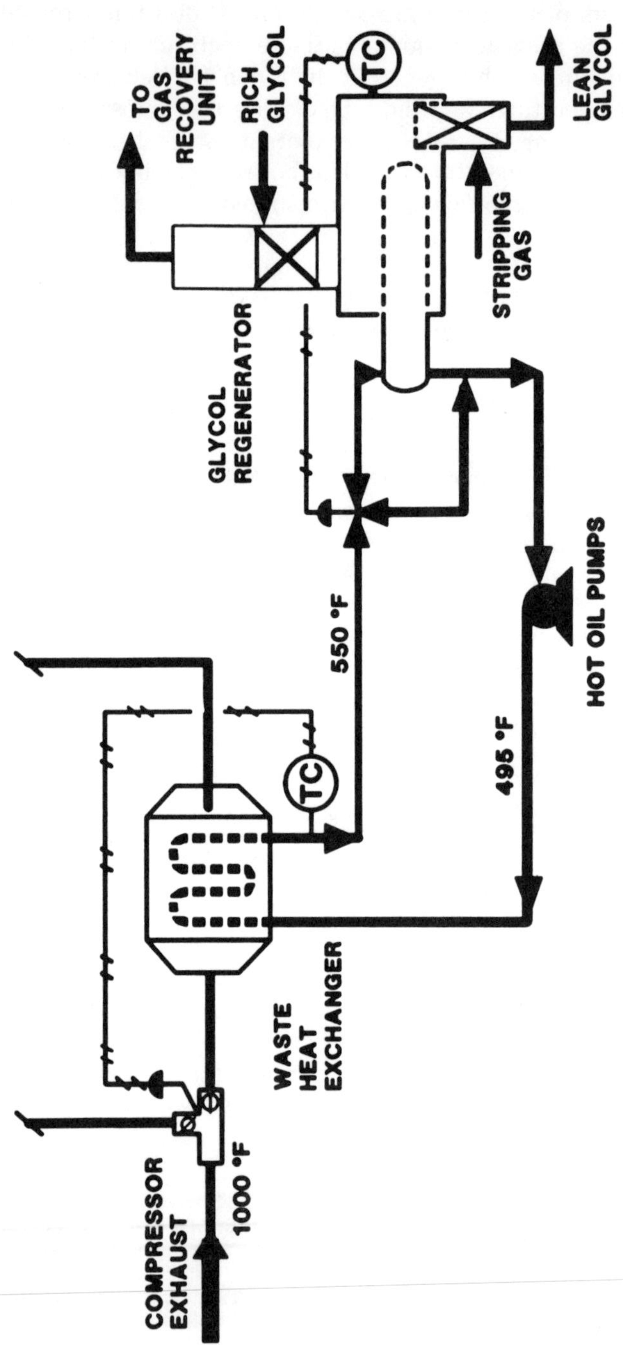



*Figure 31.* Flow diagram of glycol regeneration from engine waste heat

*Reference:* American Gas Association 1981 Operating Section Proceedings, *Catalog Number X50681; "Waste Heat Recovery from Reciprocating Gas Engines: Glycol Regeneration," Page T-26*

There are many factors that should be considered in the installation of heat recovery equipment because it ultimately affects the operation of the gas engine or turbine. The waste-heat boiler and ducting installation must be sized to prevent excessive back-pressure on the engine. Where waste heat is used in a process requiring continuous supply, a fuel-fired standby boiler must be available when maintenance is required on the engine/compressor, or when there is no demand for the compressor.

Internal combustion engines have lower heat rates than turbines and operate at air/fuel ratios close to stoichiometric, resulting in much lower exhaust mass flow rates than those in turbines with comparable ratings. The exhaust temperature at the engine typically is about 985°F (530°C), which can be lowered to about 325°F (162°C) at full load if the possibility of water vapor condensation in the exhaust is to be eliminated. The exact amount of recoverable heat can be calculated using Equation 18.

$$Q_H = q_m \times c_p \times (t_1 - t_2)$$ 

Equation 18

Where: $Q_H$ = Heat recovery in desired units, Btu/h or J/h
  $q_m$ = Exhaust mass flow rate in lb/h or kg/h
    [lb/h=ft³/h×41.13/($T$) and kg/h=m³/h×20 940/($T$),
    where $T$ is the absolute temperature at which the
    volumetric flow rate was determined]
  $c_p$ = Average specific heat capacity of the exhaust gas
    [0.258 Btu/(lb•°R) or 1.08 kJ/(kg•K)]
  $(t_1 - t_2)$ = Exhaust temperature drop due to heat recovery °F
    (°R) [°C (K)]

The heat recovered from the exhaust under the conditions outlined above represents about one-half of the 30 percent of the fuel input value rejected into the exhaust. Another 30 percent normally is rejected into the water jacket system, enhanced by water-cooled manifolds found on most gas-fueled engines.

The economics of using waste energy indicates a quick payout where a reduction of fuel gas can be realized and/or where the energy can be sold or utilized for additional plant needs.


## Exhaust Silencing

Noise created by the exhaust gas leaving the engine generally is above the acceptable level, and silencing must be provided. Silencing can be accomplished by the installation of an effective silencer, as illustrated in Figure 32 and in the cross-sectional view shown in Figure 33, and the proper sizing of the exhaust piping from the engine to the

*Figure 32.* Engine exhaust silencer and combustion air filter

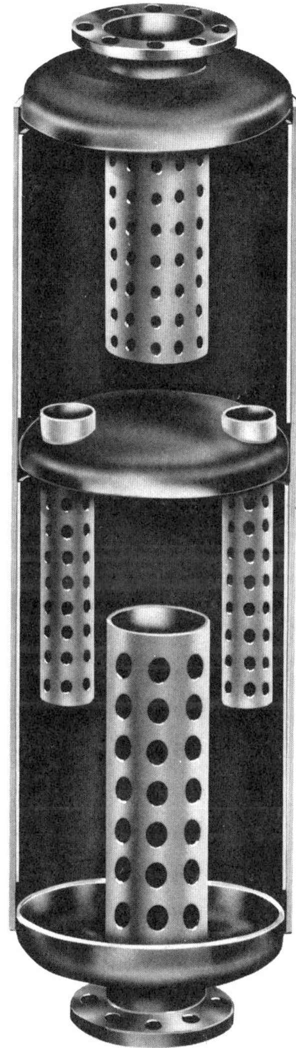

**Figure 33. Cross-section of typical engine exhaust silencer**

silencer. The silencer and piping must reduce not only the noise level, but also must not impose excessive back-pressure on the engine. Excessive back-pressure results in power loss, poor fuel economy, and excessive valve temperatures. To determine the type of silencer to use, the desired noise level must be established at some distance from the noise source.

Since no silencer is 100 percent effective, consideration should be given so that the type of silencer selected will not violate the plant noise acceptability criteria. This criteria must account for both the environmental noise pollution (the annoyance value to the neighborhood) and the hearing protection of the employees. Many local codes control the first by establishing a certain set of day/night exposures to noise generally measured at the property line of the facility. Note that this is a limit on the total noise at that location and not on the noise produced by a single-plant component. Therefore, if over-specification is to be avoided, it is important to evaluate the noise contributions by all sources before the proper type of exhaust silencer is recommended.* However, while the most important aspect of a noise problem is its power level, the nature of the source has to be considered as well. For instance, the blowdown of a compressor or a pipe section unexpectedly, or at night, can create problems which cannot be described totally in terms of sound level alone.

Intermittent and random operation of a device can increase its annoyance potential. Similarly, if a source has an unusual frequency distribution or pure tone components which fall into the speech interference area or areas of maximum hearing sensitivity, adverse effects will be magnified many times.


## Exhaust Piping

Exhaust piping must be sized to avoid exceeding the allowable pressure drop specified for the engine or turbine. Generally the size will be the same as, but never less than, the exhaust connection on the engine. The distance between the engine and silencer should be minimal with as few bends as possible. A high temperature, flexible connection is installed between the engine and exhaust piping. Exhaust system pressure drops should be calculated just as they were for the air inlet system.

---

*Bolt Beranek and Newman Inc., *Noise Control for Reciprocating and Turbine Engines Driven by Natural Gas and Liquid Fuel* (New York [now Arlington, VA], December 1969).

Provisions should be made to allow for thermal expansion of approximately 0.8 percent of pipe in the entire exhaust system.

## LUBE OIL SYSTEM

The lube oil system provides the means of supplying an adequate flow of clean lubricating oil to the moving machinery parts at a controlled temperature. Figure 34 illustrates a typical schematic diagram of a lube oil system for a compressor engine.

### Oil Filtration

As the lube oil in a gas engine, compressor, or gas turbine is circulated, it picks up and carries particulate matter that may be abrasive to the bearing surfaces and cylinder walls. Therefore, the oil is passed through a lube oil filter to remove these particulate solids. Usually this filter is installed in the flow line between the oil cooler and the engine oil header. The type of filter used depends upon the oil used and the required degree of filtration. A filter can be operated as a full-flow filter handling the total oil flow, or as a bypass filter handling only a fraction of the total oil flow. Generally the full-flow mode is used; however, where the removal of acids is required, bypass filtration with a filter media of Fuller's earth may be used.

Filter manufacturers furnish several types of filter elements with different filtration capabilities. Elements that remove particles as small as 2 micrometres (microns) in diameter are available. The size and number of elements used is based on the oil flow rate and the desired initial pressure drop across the filter at operating conditions.

**NOTE:** *Used oil filters must be treated and disposed of as hazardous waste.*

During cold starts the high viscosity of the oil could lead to an excessive pressure drop across the main filter, creating a lube oil supply shortage. In this case the oil can be routed to the lube oil header through a bypass relief valve and through a very coarse mesh (200-micrometre range) bypass filter. The relief valve as shown in the diagram in Figure 34, labeled as a pressure regulating value, is set at some desired maximum pressure to protect the oil pump and the main oil filter. On larger engines the relief valve will dump the oil back into the engine crankcase.

Factors to be considered in the selection of an oil filter are: the type of oil, non-detergent (ashless) or detergent (additive); oil flow rate; particle size removal; access to filter elements for inspection and replacement; and compressor equipment manufacturers' recommendations.

### Oil Temperature Control

Lubricating oils have a normal operating temperature range at which they perform best. Since oil is more viscous at low temperatures and less viscous as the temperature increases, its temperature must be controlled to maintain a desirable viscosity to allow it to function properly.

As oil circulates through machinery, it serves as a coolant as well as a lubricant. As a coolant it picks up heat generated by the bearing surfaces and from the cylinder walls of an engine. This heat then must be removed to maintain the recommended oil temperature. When an engine operates with coolant temperatures above 220°F (104°C), the heat always is rejected to a cooling medium other than the jacket water, and may be recovered to preheat the boiler feed water or to meet the domestic hot water requirements. The rate of heat rejection from a gas engine is about 330 Btu/hph (130 W/kW). A shell-and-tube heat exchanger, as illustrated in the diagram in Figure 34 and as shown in Figure 35, frequently is used for this purpose. Oil flows through the shell portion while a coolant flows through the tubes and absorbs heat from the oil. Then the heat absorbed by the coolant is removed by a coolant-to-air or a coolant-to-water heat exchanger.

To maintain a relatively constant oil temperature, a bypass control valve usually is installed in the oil piping to regulate the rate of oil flow through the heat exchanger. The temperature of the crankcase oil usually is kept below 145°F (63°C), and the heat load is about 300-400 Btu/hph (118-157 W/kW).

### Pre-Lube/Post-Lube System

After an engine has been shut down, the lube oil immediately begins to drain back into the crankcase. If the engine is down for a prolonged period of time, the bearing surfaces can become very dry, and portions of the oil supply piping and passages also can become filled with air.

To protect the engine from lack of lubrication on start-up, a pre-lube pump is incorporated to supply oil to the engine bearing surfaces prior to start-up. Before attempting to start the engine, this pre-lube cycle should have sufficient time to force all air from the system, and to provide a good film of lubricant to all bearing surfaces. The pre-lube pump is shut down after the main engine oil pump has raised the oil pressure to the normal operating pressure.

The post-lube system provides a flow of oil to the bearing surfaces during the cool-down after the engine has shut down. Because of their high operating temperatures, heavy gas turbines require a long cool-down period, and the post-lube cycle is extremely important.

Pre/post-lube pumps are electric-motor or pneumatic driven and are

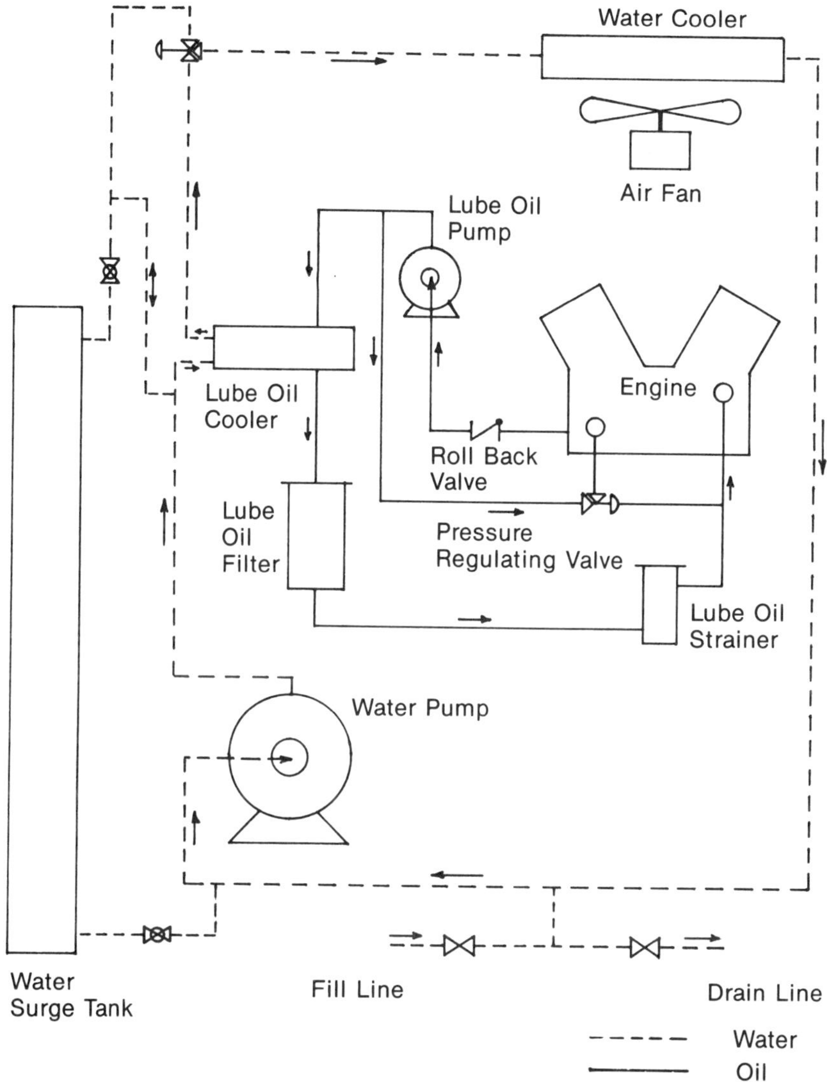


*Figure 34.* **Flow diagram of engine/compressor lube oil system**

*Figure 35.* Shell-and-tube lube oil heat exchanger showing bypass temperature control valve

automatically controlled by the start-and-stop cycle of the engine controls in most cases.

### Force-Feed Lube System

There are numerous points in a modern internal combustion engine that require different amounts of lubricant at various pressures. Because of the various conditions and requirements, some points cannot be lubricated by oil circulated by the main oil pump; consequently, lubricant to these points commonly is furnished with a force-feed mechanical lubricator. Force-feed lubricators consist of a series of plunger type pumps calibrated to deliver a measured quantity of oil to a specific lubrication point.

A modular divider valve system may be used to lubricate the many points on the engine or compressor. In this system a pump, such as the one shown in Figure 36, forces the lube oil to valve blocks, illustrated in Figure 37, which contain a series of metering pistons. Oil for this system can be supplied either from the engine crankcase or from a separate oil reservoir.

Sensors are used to give a warning or to shut down the engine if the oil is not being delivered as required. Sensors can detect a pump failure, low oil level in the pump reservoir, and a malfunctioning divider valve block.

## JACKET WATER SYSTEM

Jacket water systems provide coolant to the engine and compressors to maintain a controlled temperature.

### Compressor Water Jackets

Cooling jackets normally are provided in the reciprocating compressor cylinders, and are designed to operate at discharge temperatures above 140°F (60°C). Coolant circulation through the jacket is recommended if the discharge temperature is between 180 and 250°F (82 and 121°C), or if the stage temperature rise exceeds 170°F (94°C). If cooling jackets are provided, circulation also is recommended when valve unloading is used as a normal method of controlling the compressor capacity. Where coolant circulation is not required, the jackets may be filled with nonfreezing liquid to ensure uniform temperature distribution within the cylinder and avoid thermal distortion.

A 15°F (8.5°C) rise in coolant temperature through the compressor jackets is about the maximum. The coolant temperature should be be-

Courtesy of United Gas Pipe Line Company

*Figure 36.* Compressor cylinder lubricator pump

Courtesy of United Gas Pipe Line Company

*Figure 37.* Divider block for cylinder lubricator system

tween 145 and 150°F (63 and 66°C) to avoid cylinder lubrication and wear problems. The heat transferred from the compressor is in the range of 250 to 700 Btu/hph (100 to 275 W/kW).

## Engine Water Jackets

Reciprocating engines similarly require continuous cooling of the power cylinders. The cooling load is in the vicinity of 3 000 Btu/hph (1 180 W/kW), or about 118 percent of the engine output rating for naturally-aspirated engines. The heat rejection by a supercharged engine is about 1 200 Btu/hph (470 W/kW), or about 47 percent of the engine output rating.

The permissable temperature rise for the coolant circulating through the engine water jacket is between 10 and 15°F (5.5 and 8.5°C).

## Jacket Water Heat Exchangers

Heat exchangers dissipate the heat absorbed by the coolant after it has flowed through the engine or compressor water jackets. Heat exchangers can be either a liquid-to-air type or liquid-to-liquid type. A liquid-to-air, forced draft heat exchanger is illustrated in Figure 38.

The exchangers should be sized to handle the full flow of the jacket water pump with a minimum pressure drop, and have sufficient sur-

Courtesy of United Gas Pipe Line Company

***Figure 38.*** **Forced draft, fin fan heat exchangers**

face to dissipate the heat picked up by the coolant at the maximum operating load conditions and ambient air temperatures. Fouling of the heat exchanger surfaces, after an extended period of usage, also should be anticipated.

Forced or induced draft fans are employed to increase the air flow through the heat exchangers thereby reducing the total required heat surface. Various methods of driving the fans can be used such as electric or hydraulic motors, gas engines, main engine jackshafts, and engine-driven electric generator drivers. Jacket water heat exchangers are similar to gas heat exchangers which are discussed in more detail below under Gas Cooling.

### Jacket Water Pumps

The water pump for an engine/compressor jacket water system circulates the cooling water through the heat exchangers, and generally is driven directly from the engine crankshaft, as shown in Figure 39; from a power takeoff shaft, or by a belt drive system. Larger engines may have the pump mounted away from the engine and driven independently by other means such as an electric or hydraulic motor. The

*Courtesy of United Gas Pipe Line Company*

*Figure 39.* **Reciprocating engine cooling water pump**

water pump must be sized to provide adequate flow not to exceed a certain temperature rise across the engine at maximum operating load.

Depending on the nature of the cooling system, it may be advisable to have an auxiliary pump to provide additional circulation during heavy loading. In cases where it is desirable to have a separate cooling system for lube oil cooling or for compressor cylinder cooling, an additional pump and heat exchanger will be required, as illustrated in the schematic diagram in Figure 40.

## Jacket Water System Coolant

Coolant in the jacket water system may be either raw or treated water, but often an antifreeze solution is used. The efficiency of the heat transfer system is dependent primarily upon the type of coolant flowing through it. Whether using strictly water or a water/antifreeze mixture, good quality water should be used. An antifreeze solution always should be used for cold weather operations.

## Jacket Water Corrosion Control

Water available for use in the jacket water system of an engine may have impurities and be of a chemical compositon that will cause the formation of scale and sludge, as well as set up corrosive action in the engine, piping, and heat exchangers. Poor quality water without proper treatment will result in poor heat transfer and could damage the equipment; consequently, a good jacket water corrosion control program should be applied to the jacket water system.

## Jacket Water Temperature Control

Because of ambient temperature changes and variations in the engine load conditions, the temperature of the jacket water can fluctuate widely and have an adverse effect on the operation of the engine. Water temperatures therefore should be controlled in accordance with the engine manufacturer's recommendations. This commonly is done with a three-way, thermostatic control valve installed in the jacket water piping. This valve senses the temperature of the coolant as it leaves the engine and regulates the amount of coolant flowing to the heat exchangers. Variable speed and variable pitch fans often are used in heat exchangers to provide additional temperature control.

In the event that the control is not maintained properly, there should be a high-temperature alarm and shutdown device to protect the engine from damage resulting from excessive jacket water temperature.

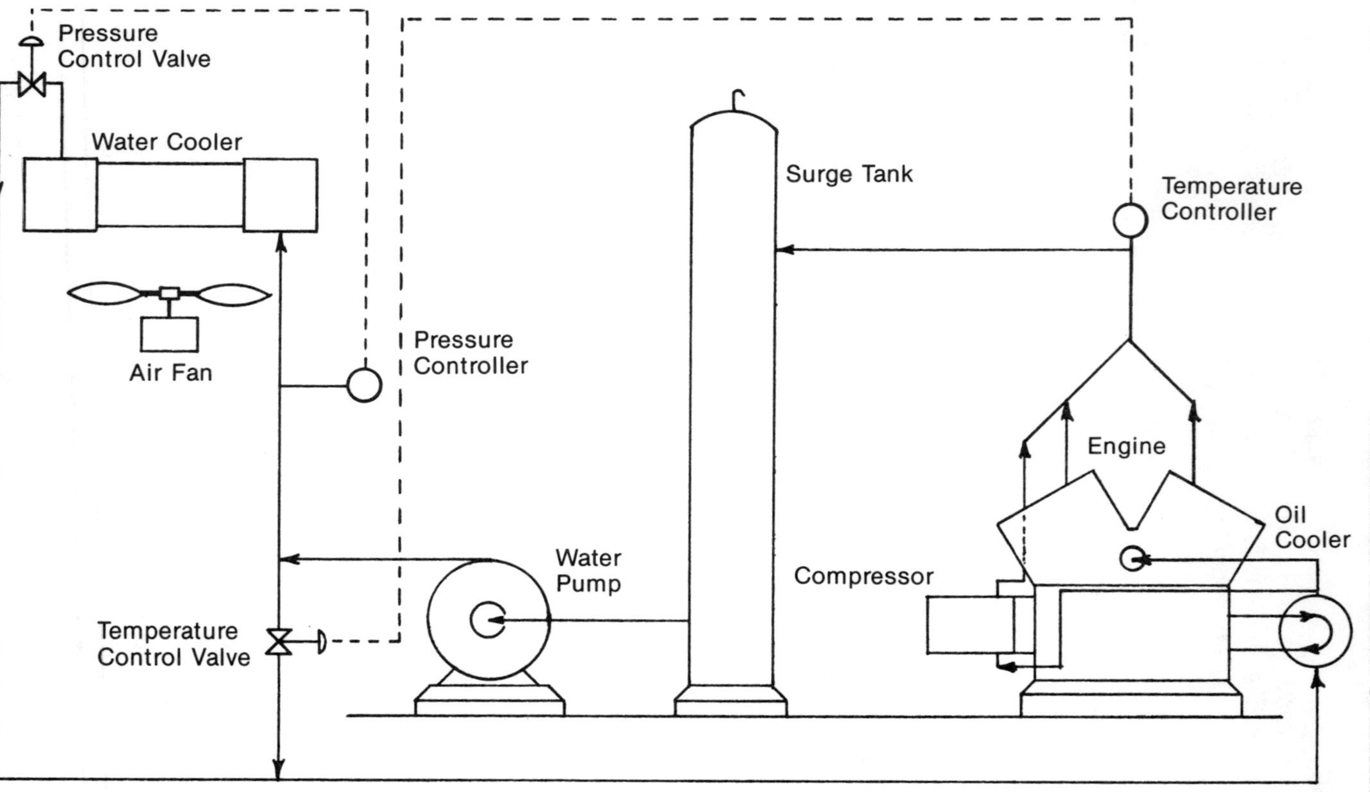


*Figure 40.* Flow diagram of engine/compressor jacket water system

## STARTING SYSTEM

There are several methods of starting internal combustion engines and turbines, but any starting system should be capable of achieving high break-away torque to start the equipment rotating, and then drive the engine or turbine to a self-sustaining speed until firing has occurred. The three types of starting systems most commonly used are the air-injection type, expander type, and auxiliary engine type.

### Air-Injection Starters

For large integral gas engine/compressor units, an air-injection system is the most common starting method. This type of starting is accomplished by the injection of high-pressure air at 250 lbf/in² (1 720 kPa) into the power cylinders in a predetermined sequence by cam-operated valves to bring the engine to starting speed.

### Expander Motor Starters

Expander motors commonly are used for cranking smaller engines, utilizing air or gas pressure of 150 lbf/in² (1 030 kPa). The expansion of the air or gas through the motor provides energy to rotate the engine. Depending upon the required torque, single or multiple-motors are used, as shown in Figure 41. These motors also can be used in starting systems of gas turbine units.

### Engine and Turbine Starters

Although actual starting power requirements can be determined only through detailed engineering, single-shaft gas turbines normally require about 10 percent of the power output rating as input during start-up, and about 3 percent for two-shaft gas turbines. Diesel engines, electric motors, and steam turbines are used occasionally as starters for large gas turbines. A clutch engages the starter on start-up, and automatically disengages when the gas turbine reaches self-sustaining speed. Note that a steam turbine used for starting does not have to be declutched after the gas turbine has reached self-sustaining speed. It can be used as a driver's "helper" in situations when the driven load exceeds the available gas turbine power. When not needed, it can freewheel with cooling steam injection.

## FUEL GAS SYSTEM

The engine fuel gas system must be designed to furnish clean, dry

*Figure 41.* **Expander starter motors for reciprocating engine**

gas to the engines or turbines at the required pressure and flow rate. Ordinarily the fuel gas supply is taken from the main gas headers at the compressor station. The following are the major components that make up the fuel gas system.

## Fuel Gas Pressure Regulators

Fuel gas regulators must be installed to reduce the high pressure gas supply to the required pressure at the engines or turbines. When the pressure at the supply point is much higher than that required, more than one stage of regulation may be necessary.

## Fuel Gas Meters

The desired flow rate of the fuel gas system will determine the type and size of meter station for measuring the fuel gas. Measurement can be accomplished with positive displacement meters, rotary meters, or orifice meters. Frequently individual meters are installed at each engine to provide operating performance data as well as automatic load control for the compressors.

## Fuel Gas Separator

To ensure clean, dry fuel gas, a fuel gas filter/separator, as shown in Figure 42, may be required upstream of the fuel gas measuring and regulating facilities. The piping for the separator should include a bypass arrangement to allow for filter element replacement without shutting off the fuel flow. Some installations may require an automatic dump valve to evacuate liquids trapped in the separator.

## Fuel Gas Heater

Sometimes it is necessary to install a fuel gas heater to heat the gas before it has expanded through the pressure regulators to prevent hydrate formation in the piping and regulators. The heater is generally a shell-and-tube type with the heat being supplied from either the engine jacket water system or a gas-fired boiler. Bypass piping should be provided in cases where the heater is not required continuously.

*Courtesy of United Gas Pipe Line Company*

*Figure 42.* **Engine fuel gas filter/separator**

### Fuel Gas Receiver

A fuel gas receiver may be installed near the engines to eliminate fluctuations in the gas pressure caused by pulsations of the engine, and variations in the pressure and flow rate from the regulator station.

## VENTS AND DRAINS

Vents are necessary to relieve pressure and liquid accumulations, that could prove hazardous during operation of the equipment.

### Compressor Packing Gland Vent

Each compressor packing gland must be vented to prevent gas that might leak past the packing from entering the engine crankcase or leaking into the compressor building. This vent should be piped to the outside of the building and to an area where the vented gas will not create a hazard.

### Jacket Water Vent

A vent should be provided in the jacket water expansion (surge) tank to allow the tank to operate at atmospheric pressure. The expansion tank provides a high point in the system for venting. Other high points that could act as traps for air also should be vented.

### Crankcase Vent

Crankcase vents must be provided on internal combustion engines to allow accumulated gas buildup to escape rather than create a hazardous combustible mixture in the crankcase. This vent is piped to an area outside the building to a point where any oil-entrained vapors could not be drawn into the engine air-intake system.

### Oil Tank Vent

Gas turbines often have an oil reservoir built into the skid package, which must be vented to prevent any pressure buildup within the tank. These vents also should be piped to a safe location outside the building and away from the air intake. The installation of a flame arrestor also should be considered to reduce the possibility of fire migrating to the oil tank in the event of a fire at the vent outlet.

## Compressor Crosshead Oil Drain

Compressor crosshead oil drains are necessary to convey any excess oil wiped from the compressor rod by the rod wiper packing to a waste oil tank.

## Oil and Water Drains

Piping should be provided to drain the oil and coolant when servicing the equipment. To drain the crankcase, usually the oil is drained from a low point in the oil piping. The coolant, whether it is treated water or an antifreeze solution, can be drained back so that it can be used again.

# ENGINE AND COMPRESSOR FOUNDATIONS

Machinery foundations are one of the most important aspects of compressor station engineering. Such factors as soil analysis, reinforced concrete design, and dynamics of foundation systems require particular attention. The technology is now available to investigate analytically such important phenomena as resonant frequencies and dynamic instability of machinery foundations.

## FOUNDATION DESIGN CONSIDERATIONS

Since foundation design is a highly technical science, a competent foundations engineering staff should be consulted prior to starting any major compressor station project. Only a brief introduction of the most common design considerations will be discussed in this text.

There are two basic types of foundations used for compressor units in the gas industry. The first is a simple concrete pad similar to the two examples shown in Figure 43. It is used for most small skid-mounted reciprocating units and for high-speed centrifugal compressors. The second type is illustrated in Figure 44, and it incorporates a combination of reinforced block and mat concepts. The block and mat design is common for large, low- or medium-speed reciprocating compressors. The block provides sufficient mass to ensure stability for the dynamic load characteristics of the machine, and the mat provides the required load distribution so that the allowable soil-bearing capacity is not exceeded.

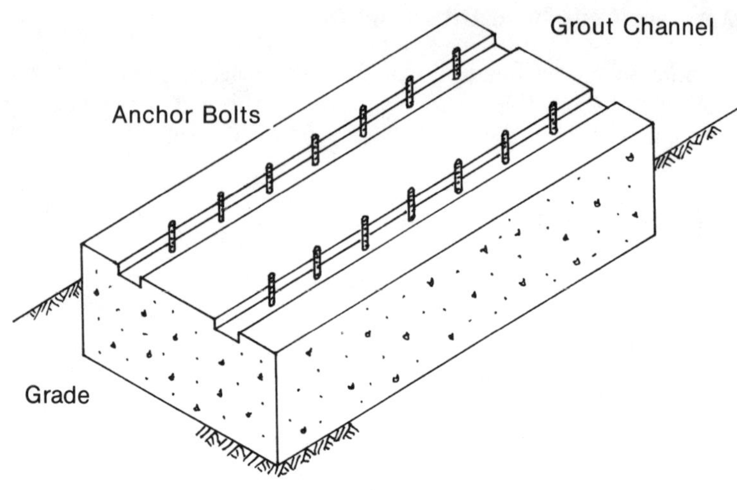


a. Continuous Pad Foundation

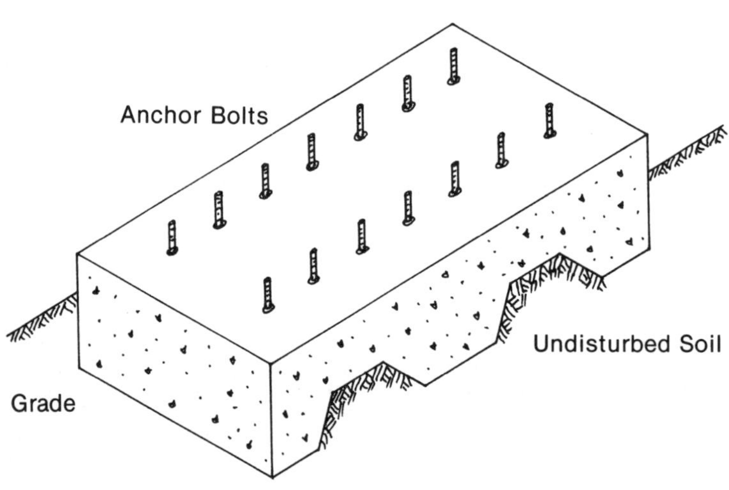


b. Waffle Type Pad Foundation

**Figure 43. Typical concrete pad compressor foundations**

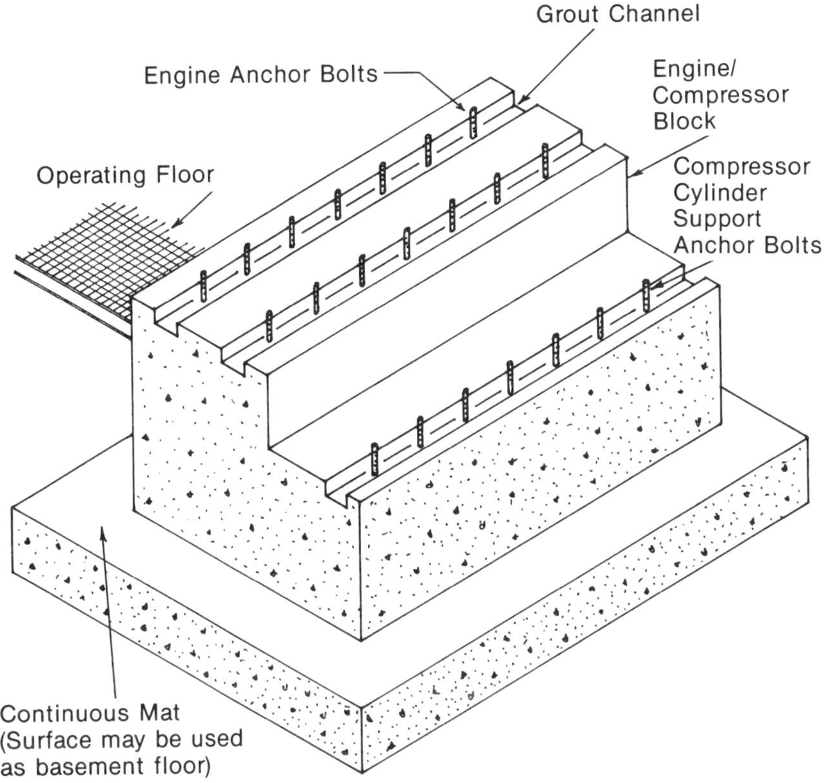


*Figure 44.* **Concrete block and mat compressor foundation**

## Soil-Bearing Considerations

A complete soil study is the first task in designing a foundation for compressor units. Dynamic loading imposes special problems on the behavior of soils that may not occur under static conditions. Consequently, it is extremely important to obtain soil mechanics information regarding soil type and load-bearing capacity, consolidation and settlement characteristics, and the resonant frequency of the soil strata.

In some cases, particularly where loose sands or saturated silts and clays are encountered, it may be necessary to use driven steel or wooden piles rather than risk the problems experienced with excessive foundation settlement. Another method in lieu of driven piles is to incorporate soil stabilization techniques using chemical or cement grout injection. Where soils are weak, drilled-in-place piles or underreamed footings should be considered, particularly where a weak upper strata is underlaid by a strong strata.

Driven piles also are used to reduce foundation vibrations since the piles increase the effective volume of the soil system. Although the piles increase the effective soil mass, it is important to recognize the fact that the load of the foundation and machinery is supported by the piles since the soil immediately below the foundation will settle away from the foundation base.

Soil mechanics is a highly specialized science, and extra effort in obtaining competent consultation in determining the nature of the load-bearing material for large machinery is well worth the investment. It is extremely expensive, and often impossible, to correct problems that result from foundation instability or continuous settlement. Entire compressor stations have been relocated simply because the nature of the soil had not been evaluated properly during the original design phase.

Once a preliminary foundation has been designed, the total weight of the machinery and foundation may be calculated, and the resulting soil-bearing pressure in lbf/ft$^2$ (kN/m$^2$) can be determined and compared with the permissible soil-bearing capacity for the specific soil characteristics. Soil-bearing capacity determination is complex, and should be made by analytical methods as given in various references on the subject of soil mechanics. Presumptive soil-bearing pressures are those that have been found from general experience, and often are used for simple approximations in the design of foundations. Typical values of presumptive soil-bearing pressures are given in Table 1. These values may be used when there is reasonable assurance that the soil characteristics are well-known, and the soil behavior has been proven by experience with other neighboring structures. This method does not preclude the soil-boring analysis, which should be done regardless of the conditions found in adjacent construction projects.


## Foundation Design

Before attempting to design a foundation for a large compressor unit, certain information needs to be obtained from the manufacturer. The following minimum data should be available to the foundation designer:
- Total weight of the package
- Base dimensions of the package
- Operating speed of the machine
- Radius of the crankshaft throws

## TABLE 1
### Typical Presumptive Soil-bearing Pressures for the Preliminary Design of Foundations

| Material | kips/ft² | kN/m² |
|---|---|---|
| Loose sand, dry | 1.5–3 | 70–140 |
| Firm sand, dry | 3–6 | 150–300 |
| Dense sand, dry | 8–12 | 400–600 |
| Loose sand, inundated | 0.8–1.6 | 40–80 |
| Firm sand, inundated | 1.6–3.5 | 80–170 |
| Dense sand, inundated | 5+ | 240+ |
| Soft clay | 0.6–1.2 | 30–60 |
| Firm clay | 1.5–2.5 | 70–120 |
| Stiff clay | 3–4.5 | 150–200 |
| Hard clay | 8+ | 400+ |
| Loose mica, silty sand | 2.5–4.5 | 120–200 |
| Firm mica, silty sand | 4.5–7.5 | 200–350 |
| Badly fractured or partially weathered rock | 10–15 | 500–1 200 |
| Hard rock, occasional soft seams | 30–100 | 1 500–5 000 |
| Massive hard rock | 200+ | 10 000+ |

Note: 1 kip/ft² = 1 000 lbf/ft²

(From G. F. Sowers. *Introductory Soil Mechanics and Foundations: Geotechnical Engineering,* Fourth Edition, by George F. Sowers, New York: Macmillan Publishing Co., 1979.)

- Weight of the reciprocating parts (piston, crosshead, connecting rods, etc.)
- Length of connecting rod
- Distance required to remove engine and compressor pistons and connecting rods
- Weight of unbalanced rotating parts

The base dimensions of the concrete pad or block should be such that there is 6 to 12 inches of concrete beyond all sides of the base of the unit. Manufacturers have recommended from 0.091 to 0.141 yd³ of concrete per engine horsepower (0.093 to 0.144 m³/kW) for low-speed machines.* Early researchers found that the ratio of the foundation weight to the engine/compressor weight was a suitable method of determining the size of the foundation.

---

*Ernst Rausch, *Maschinenfundamente und andere dynamische Bauaufgaben,* 2nd ed. (Berlin, 1943), 3 vols.

Manufacturers frequently recommend a foundation weight that is two to five times the weight of the machine. This information commonly has been determined as a result of successful experience with the specific machine. For large machines the concrete block may be supported by a concrete mat as shown in Figure 44 to provide adequate load distribution.

Once a preliminary foundation design has been made, it is necessary to determine the combined resonant frequency of the machine, foundation, and soil system. The natural frequency of the system may be found using the following equation:

$$f_\mathrm{n} = \frac{1}{2\pi} \sqrt{\frac{K \times g}{W_\mathrm{f} + W_\mathrm{s}}} \qquad \text{Equation 19}$$

Where: $f_\mathrm{n}$ = Natural frequency of the foundation and soil system, Hz

$K$ = Spring constant of soil mass, lbf/ft (kN/m)

$W_\mathrm{f}$ = Weight of machine and foundation, lbf (kN)

$W_\mathrm{s}$ = Weight of soil system, lbf (kN)

$g$ = Acceleration of gravity, 32.2 ft/s$^2$ (9.8 m/s$^2$)

The spring constant, $K$, is a function of Poisson's ratio, shear modulus, and the radius of the loading area.

In general the natural frequency of the system should be at least twice the frequency of the applied forces for low-speed units, or less than one-half of the resonant frequency for machines that operate at a frequency greater than the natural frequency of the system.

The natural frequency of the foundation system can be decreased by:

* Increasing the weight of the foundation
* Decreasing the base area of the foundation
* Reducing the shear modulus of the soil
* Placing the foundation deeper into the soil

The next step in foundation design is to determine the amplitude of vibration of the foundation system. This is particularly important in situations where ground-transmitted vibration could become an annoyance to neighboring residents or business occupants. Various foundation engineering textbooks cover the principles involved in analyzing the amplitude of vibration for dynamic systems. Design criteria and limiting values of amplitude with respect to frequency also are available.*

---

*G. A. Leonards, ed., *Foundations Engineering* (New York, 1962).

## ALIGNMENT AND GROUTING

Engine/compressor alignment is a highly specialized field only recently fully appreciated. The realization that broken crankshafts and other catastrophic engine failures can result from bad alignment has made clear the crucial importance of alignment and grouting of machinery.

Heavy machinery is mounted on a concrete foundation using either a baseplate (full-bed) or a soleplate. The baseplate is a rectangular steel frame upon which the machinery and some of its ancillaries are mounted at the factory. The entire baseplate then is bolted to the foundation and grouted in place as shown in Figure 45. Baseplates increase the cost of equipment, but generally pay for themselves by keeping the machinery aligned during installation. The soleplate method, as illustrated in Figure 46, incorporates small steel pads supporting the feet of individual machines, bolted and grouted to the foundation separately. They are used on all large reciprocating engines that require separate handling because of their size. The use of soleplates requires the careful alignment of equipment at the site, and may result in higher installation costs.

*Courtesy of Energy Services Group of Cooper Industries*

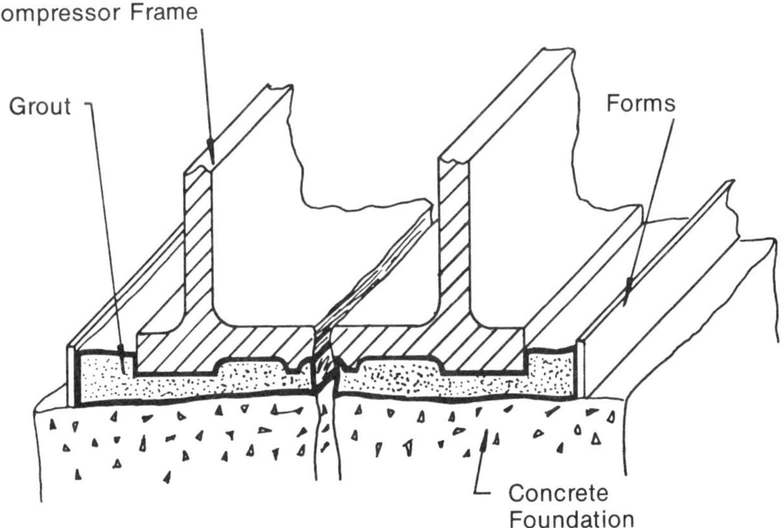


*Figure 45.* **Direct full-bed grouting method for baseplate engine/compressor unit**

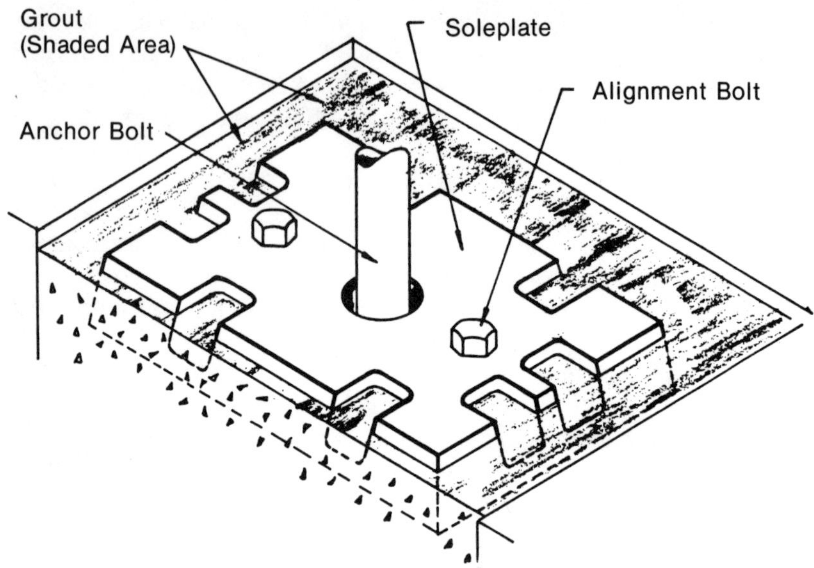


a. Top View of Soleplate Method of Grouting

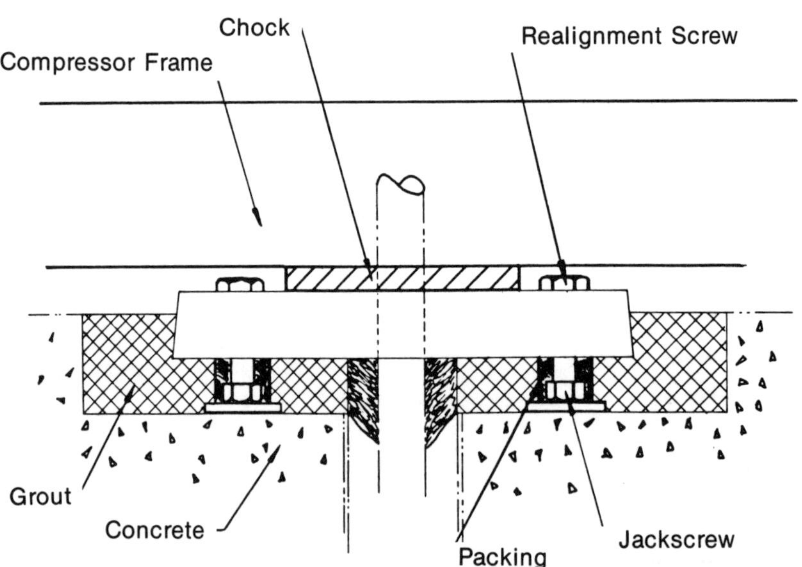


b. Side View of Soleplate Method of Grouting

**Figure 46.** Method of grouting engine/compressor units using individual chocks and soleplates

Proper level and alignment of compressors cannot be overemphasized. All anchor bolts should be tightened against jack bolts or wedges before grouting. Grout should be of a non-shrinking epoxy type. This work should be done only under the supervision of an experienced millwright or engine erector. The manufacturer's recommendations as to soleplates or other methods of controlling expansion are important. Extensive damage can result on large engines and compressors by restraining certain natural movements.

Following is a discussion of alignment and grouting techniques adopted by the gas industry.

## Engine and Compressor Alignment

For machinery to operate properly, it must be aligned correctly. The moving parts must be free to move without undue restraint. To be aligned properly, the frame of a reciprocating engine must be straight, and the crankshaft must be free of unnatural restraint when positioned in the frame. Engine manufacturers provide machined surface areas along the engine frame to use as alignment reference points. Devices used to align the frame include machinist's levels, wire lines, optical levels, straight edges, and laser beams. The final alignment check should be made with a deflection gauge inserted at each crankshaft throw. The manufacturer usually specifies maximum limits for crankshaft deflections, but on the original installation these readings should be as close to zero as possible.

## Engine and Compressor Grouting

Once the engine has been aligned, it is grouted into the foundation. Grouting maintains the alignment and holds the engine in place in spite of dynamically unbalanced forces while in operation. One of the most common methods of grouting consists of the use of soleplates and chocks, commonly known as the rail-mount method. It is frequently considered superior to full-bed grouting.

Rail-mounting employs a soleplate and chock system at each anchor bolt, as shown in Figure 46. The soleplate is embedded into the concrete block and is inserted around the anchor bolt by means of a U-shaped slot. The soleplate supports a chock, which is either steel or epoxy, and also fits around the anchor bolt. The weight of the engine is transferred from the chock through the soleplate and onto the concrete block. Recently there have been installations where soleplates were eliminated in favor of a high strength epoxy base.

Full-bed grouting, as shown in Figure 45, is accomplished by leveling the compressor frame approximately 2 inches (50 mm) above the rough concrete block and, after properly positioning forms to contain the grout, pouring the semi-liquid grout material beneath the entire compressor frame. This process may be done with either sand/cement grout or epoxy grout. However, this method of grouting engines is becoming obsolete because it is nearly impossible to make an alignment correction without removing the entire grout bed and repeating the process from the start. In addition, full-bed epoxy grout is extremely difficult to remove from the compressor frame.

One method of realignment of compressor units that have been placed on full-bed grout is an injection grouting process. Injection grouting is done by drilling holes through the engine/compressor baseplate and injecting under high pressure a liquid grout material in an effort to raise the compressor unit until alignment corrections have been made. Although this method has been somewhat successful, there is a greater risk involved in attaining the desired alignment than with the soleplate method.

One advantage of the soleplate system over the full-bed grouting method is that the engine is positioned about 1 inch (25 millimetres) above the foundation. This elevation allows the circulation of air through the void between the bottom of the main frame and the foundation. Another advantage, especially where epoxy chocks are used, is that alignment adjustments can be accomplished by replacing the chocks as opposed to the more difficult, injection grouting process.

The primary disadvantage of using steel chocks is that it is often difficult, if not impossible, to achieve total contact between the chock, the soleplate, and the main frame bottom. The contact dilemma is due partially to the difficulty of installing each soleplate perfectly parallel to the bottom of the main frame. This is complicated also by the fact that on all machined metal surfaces, high and low points exist, even if they appear smooth. Consequently, the high points must support the load. As vibration causes the high points to wear away, the chocks become loose and misalignment occurs. The epoxy chock system minimizes the contact problems.

One good feature of an epoxy chock system is that the chock is installed in a liquid (viscous) form. As a liquid the epoxy is capable of completely filling the space that the chock will occupy, and near total contact is achieved—overcoming minor misalignment between the soleplate and the bottom of the main frame. A disadvantage of the epoxy chock system is the relatively low compressive yield strength of the epoxy. Care must be taken to ensure that the chock is of adequate size to support the engine once the anchor bolts have been tightened to the final torque. Since a maximum temperature of 200°F (93°C) is recommended for epoxy chocks, the type of service temperature may prohibit their use.

### Installation of Rail-Mounts on Existing Units

When regrouting an existing compressor unit, it is necessary to first remove the original grout material. If the unit is situated inside a building, it is beneficial to erect a tent around the engine to prevent fouling of the rest of the building from dust resulting from the chipping operation.

The foundation block should be chipped out at each anchor bolt to accept the soleplate. The chipping should extend to good concrete and should provide adequate clearance around the soleplate. The grouting material vendor will provide specifications for adequate clearance.

After chipping, the soleplates can be positioned and leveled at each anchor bolt as shown in Figure 46. The leveling does not have to be precise, but should be relatively close to the desired level. The soleplate should be set leaving a gap of 0.75 to 1.0 inch (20 to 25 millimetres) between the soleplate and the main frame. Forms should be installed to contain the grout used in the soleplate area. Extreme care should be taken so that the forms do not allow the grout to leak.

Next the grout to be used for supporting and containing the soleplate may be poured. The maximum dimensions of a single pour will depend on the product and the foundation temperature. Again, the vendor's recommendations should be followed.

Once the soleplate grout has cured, the engine should be leveled, and the chocks can be installed. Where epoxy chocks are used, the forming material for pouring the chock should be installed. It is advisable to spray the void with a suitable release agent to ensure that the chock does not adhere to either the soleplate or the main frame. The release agent permits the removal of the chock should future realignment be necessary. After the chocks have cured, the forms should be removed, and the anchor bolts should be tightened to the proper torque value. Then final alignment readings should be taken for future reference. Figure 47 is a photograph that shows an engine/compressor unit in which the full-bed grout was replaced by the soleplate method.

# GAS PIPING

The gas piping system in a compressor station includes the valving, pulsation control equipment, overpressure protection devices, cathodic protection facilities, and structural supports to route the gas through the compressor and gas conditioning facilities. Figure 48 shows a schematic diagram of the piping system at a small compressor station. Figure 49 gives a general flow diagram for a storage/withdrawal compressor operation. A major compressor station pipe yard is shown in the photograph in Figure 50.

*Figure 47.* **Existing compressor unit re-grouted with steel chocks and soleplates**

## GAS PIPING DESIGN CRITERIA

In designing the piping system the following items must be taken into consideration: operating pressures and capacity, pressure drop, gas velocity, mechanical strength and composition of pipe materials (grade and wall thickness), temperatures, effects of corrosion, safety, efficiency of operations, and economy of installation.

### Operating Capacity and Pressure

Design operating capacity and pressure in a piping system are based on the expected peak flow, historical data, minimum upstream and maximum downstream pressure requirements, and the pressure rating of the pipeline receiving the gas.

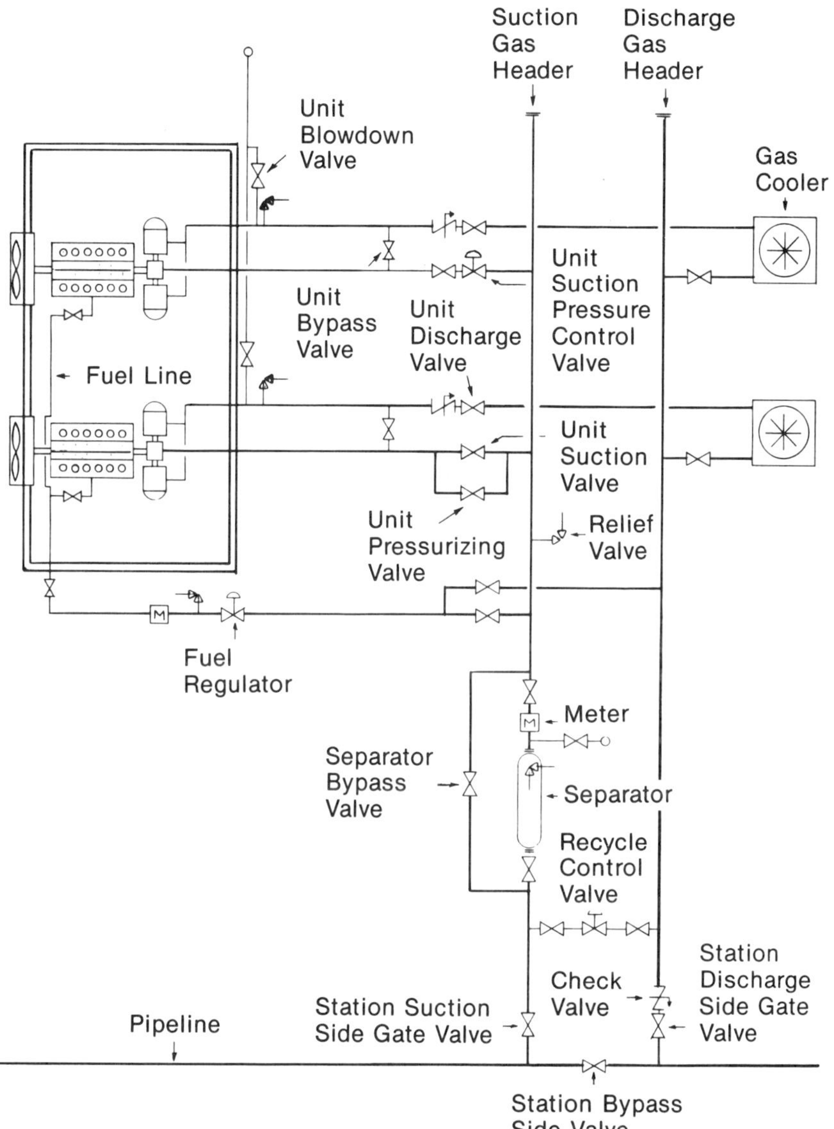


**Figure 48.** Typical reciprocating compressor station piping schematic illustrating valve applications

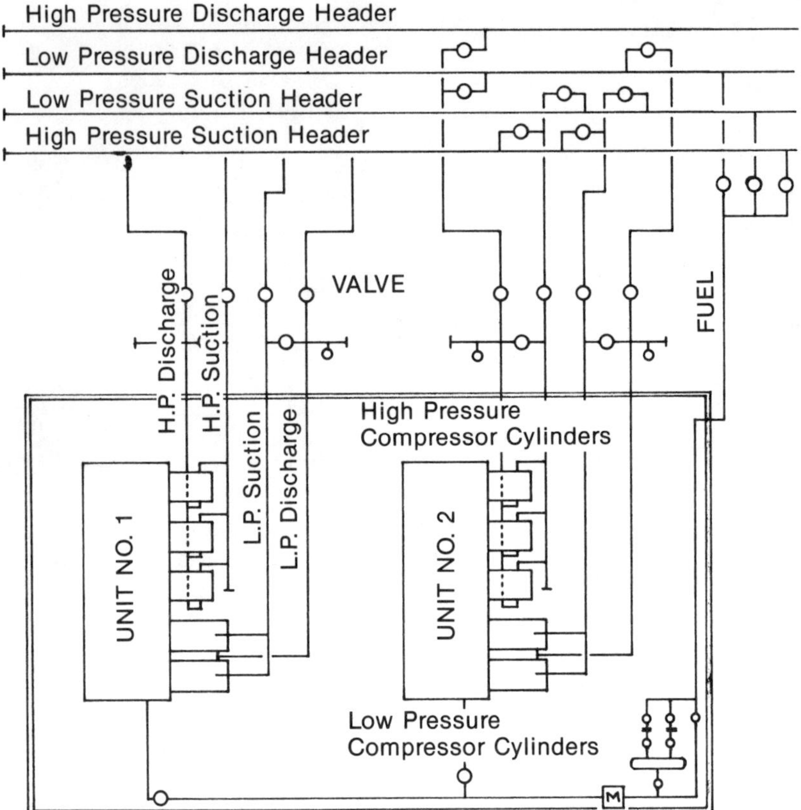


*Figure 49.* **Compressor unit gas flow diagram for storage injection and withdrawal service**

## Pipe Sizing

To determine pipe diameters, the acceptable pressure drop and gas velocity within the compressor piping must be known. Pressure drops in compressor station piping result from energy losses caused by friction between the gas and pipe wall and obstructions to gas flow. In general, energy losses should not exceed 1 to 2 percent of the total station power requirements. Power loss can be predicted by calculating station horsepower requirements at design flow rates with and without considering friction losses. Pressure drops that cause an excess of 1 to 2 percent power loss are acceptable for low-pressure systems [discharge

*Figure 50.* **Typical compressor station pipe yard**

pressures less than 300 lbf/in² (2 070 kPa)], but should not exceed 5 percent of the total pipeline pressure.

Gas velocities in piping should not exceed 2 000 ft/min (10.2 m/s) since velocities greater than this value generate an excessively turbulent flow. The velocity can be determined using the following equation:

$$v = \frac{127.3 \times 10^3 \times Q \times p_b \times T \times Z}{D^2 \times p_m \times T_b} \ \text{ft/min} \qquad \text{Equation 20a}$$

or

$$v = \frac{14.73 \times Q \times p_b \times T \times Z}{D^2 \times p_m \times T_b} \ \text{m/s} \qquad \text{Equation 20b}$$

Where: $v$ = Velocity in desired units, ft/min or m/s
$Q$ = Volume rate of flow, MMcf/d (m³/d)
$p_b$ = Base pressure, lbf/in² (kPa) absolute
$T$ = Flowing temperature, °R (K)
$Z$ = Compressibility factor
$D$ = Diameter of pipe, in (mm)
$p_m$ = Flowing pressure, lbf/inffi (kPa) absolute
$T_b$ = Base temperature, °R (K)

## Piping Connections for Smooth Flow

Straight runs of compressor flange-size piping should precede the inlet and discharge flanges of a dynamic compressor. This will ensure uniform flow distribution into the volute (allowing operation at rated capacity and efficiency), smooth diffusion (conversion of velocity heat into pressure) without excessive pressure drop in the transition zone as the gas enters the discharge, and accuracy of pressure and temperature data obtained by measurements. For services where the main line equals or exceeds the flange size of the compressor, the straight run length should be at least three times the nominal compressor flange diameter. For services where the line size is smaller than the compressor flange, the straight run should be at least three times the line size in length. However, when size transitions are required closer than three diameters from the compressor flange, they should be made using low single-tapered transition pieces rather than abrupt swages to minimize velocity profile disturbance.

## Pipe Grade and Wall Thickness

Gas piping used in a compressor station must be manufactured in compliance with the standards referenced in the Department of Transportation (DOT) Code, Title 49, Part 192. (See Appendix A). Barlow's formula is utilized to determine the nominal wall thickness of the pipe. This formula considers the grade and yield strength; gas temperature; gas pressure; longitudinal joint and design factors; and class of location and is as follows:

$$p = \frac{2 \times S \times t}{D} \times F \times E \times T \qquad \text{Equation 21}$$

Where:  $p$ = Design pressure in desired units, lbf/in² or kPa
$S$ = Yield strength of material, lbf/in² (kPa)
$D$ = Nominal outside diameter of pipe, inches (millimetres)
$t$ = Nominal wall thickness of pipe, inches (millimetres)
F = Construction type design factor
E = Longitudinal joint factor
T = Temperature derating factor

Other factors which should be considered in piping design include external loads, vibrational and thermal stresses, corrosion, and type of pipe joints. Any of these factors may make it necessary to increase the pipe wall thickness beyond that calculated for the design pressure using only Barlow's formula.

### Piping Supports

Supports for compressor station piping are utilized where any of the following are critical:

- Thermal expansion or contraction stresses
- Thrust forces caused by internal pressure
- Shear and bending stresses generated by the weight of the pipe and valves between supports
- Vibration stresses generated by gas pulsation.

As a general rule, above-ground pipe supports should be located at the ends of pipe, at changes in pipe direction, near valves, and at other locations so as to adequately support the weight of the pipe. It is important to keep the forces and moments imposed on the compressor flanges to levels low enough so as not to disturb the coupling alignment as the piping cycles between the ambient and operating temperatures. Anchors or expansion stops may be needed near the compressor flanges. For below-ground piping, leveling blocks should be located such that thermal and/or pressure reaction stresses are controlled. Leveling blocks essentially are adjustable supports such as the one shown in Figure 51, and provide a method of adjusting the pipe support periodically to ensure that the piping system is restrained even though there is gradual shifting of the piping or support itself.

The size of the support is based on the magnitude of the force acting on the support and the allowable soil-loading condition.

## PIPING VALVES

Valves are installed in the compressor station piping to route gas, isolate equipment or piping, and vent gas when required.

### Station Bypass and Side Gate Valves

Station bypass valves are installed in the mainline. When the station is in operation, the bypass valve is closed – diverting the gas through the open suction-side gate valve, the compressor, the open discharge-side gate valve, and back into the pipeline.

The suction- and discharge-side gate valves, when closed, isolate the compressor yard piping from the mainline.

These three valves, shown in Figure 48, sometimes are referred to as "fire gates," as they may be used to isolate the station in case of fire. Usually they are full-opening valves equipped with air- or gas-powered operators, and generally placed in an accessible location for ease of operation and maintenance.

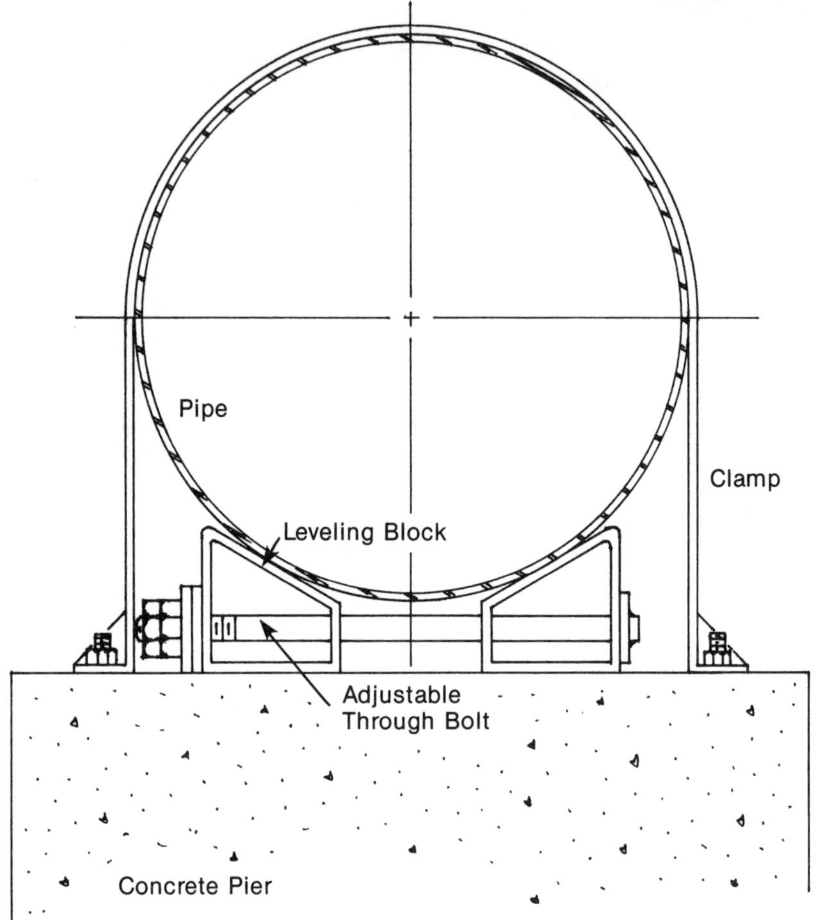


*Figure 51.* **Adjustable pipe support with leveling block**

## Compressor Unit Valves

Compressor unit valves are installed in the piping system of each compressor unit to provide a means of isolating and blowing down individual units, and loading and unloading the compressors. Ball valves equipped with power operators commonly are used for unit suction, discharge, bypass, and blowdown valves. Occasionally a relatively small pressurizing valve is used to bypass a closed suction valve to provide a method of purging and pressurizing the compressor prior to opening the larger suction valve.

## Check Valves

When two or more compressors are installed in parallel, a check valve is installed in each compressor's discharge lateral to prevent backflow. In a centrifugal compressor reverse rotation cannot be tolerated either by the compressor or the driver since in most cases the thrust and journal bearings, as well as other components, are designed to function in a forward direction only. A still greater danger is reverse overspeed since there are no brakes to limit the speed in the reverse direction. When the check valve is installed near the compressor outlet flange, it will minimize the amount of gas that could cause reverse flow during surge.

A check valve in the discharge lateral simplifies the starting and loading procedure of reciprocating compressors since the bypass valve is used as a means of loading and unloading.

In some installations a check valve is used as a station bypass valve. It allows the gas to flow automatically past the station when the compressors are shut down and the upstream and downstream pressures practically are equal.

## Miscellaneous Valves

Other valves are used to isolate separators, heat exchangers, metering facilities, and other major appurtenances. In many cases it is also desirable to install a bypass valve in addition to shutoff valves to permit the flow of gas around appurtenant equipment when it is out of service.

## PRESSURE CONTROL DEVICES

Pressure control devices are used at compressor stations to maintain gas pressures at set limits. This text will not attempt to determine the size of pressure control devices. For further information on sizing, refer to the Instrument Society of America (ISA) Standard S39.3, Control Valve Sizing Equations for Compressible Fluids, found in all valve manufacturers' sizing data sections.

## Regulators

Regulators are used to reduce high-pressure gas to a lower pressure, and to maintain a constant lower pressure. Most regulators at a compressor station are used for controlling fuel gas or instrumentation media pressure. Important elements to be considered when selecting regulators include temperature, orifice sizes, inlet pressure, outlet pressure, materials, and spring range.

### Suction Pressure Control Valves

Sometimes suction pressure control valves are used on the station inlet piping of small, unattended compressor units to protect the compressor from going down because of engine overload. They also facilitate starting of the compressor units. Their function is to prevent the suction pressure to the compressor from exceeding the engine or compressor load capability.

### Recycle Control Valves

Recycle control valves are used occasionally for controlling the low suction pressure into a compressor station, the high discharge pressure leaving a station, or both. Recycle control valves are used also on centrifugal compressors to prevent surge.

## GAS PULSATION AND CONTROL

Gas pulsations are dynamic pressure fluctuations that act as acoustic pressure waves in closed piping systems. The pulsations are produced by the sudden, periodic compression and expansion of gas within the reciprocating compressor. Pulsations result in operational inefficiencies in compressors, deviations in compressor loading and capacity, and damaging vibration in both the compressor and the station piping.

Gas pulsation can be limited by the proper design and installation of gas pulsation dampers and compressor station piping. Complete system design and analysis studies can be performed by the Southern Gas Association's Dynamic-System Analog, which determines the optimum pulsation suppression based on an acoustical analog simulation.

### Pulsating Gas Piping Design Considerations

When pulsation forces within the gas stream coincide in frequency with the mechanical natural frequency of the piping system, the resonant displacement amplification can be increased to many times the initial value.

To prevent damaging vibration, the following design considerations should be given special attention:

- Mechanical resonance of any piping segment should be at least five times greater than the maximum anticipated pulsation frequency.

- Compressor manifolds should be analyzed to determine vibration resonant frequencies.
- Vibration amplitude in any piping segment should not exceed 20 mils (0.508 millimetres). Vibration amplitude may be monitored by instrumentation.
- Judicious selection of location and type of supports, wall thickness, configuration of piping, and location of the branch connections should be made.

## Pulsation Dampers

Gas pulsation dampers are pressure vessels that provide attenuation close to the pulsation source (compressor) to protect the piping from damaging vibration. Pulsation dampers must be designed for each specific application. The size and configuration of the device is dependent upon the characteristics and severity of the pulsation. Figure 52 illustrates a common type of pulsation damper used on a reciprocating compressor.

## Allowable Pressure Pulsations

Pulsation dampers together with careful design of laterals and headers generally reduce but do not eliminate completely the shaking forces. American Petroleum Institute (API) Standard 618, *Reciprocating*

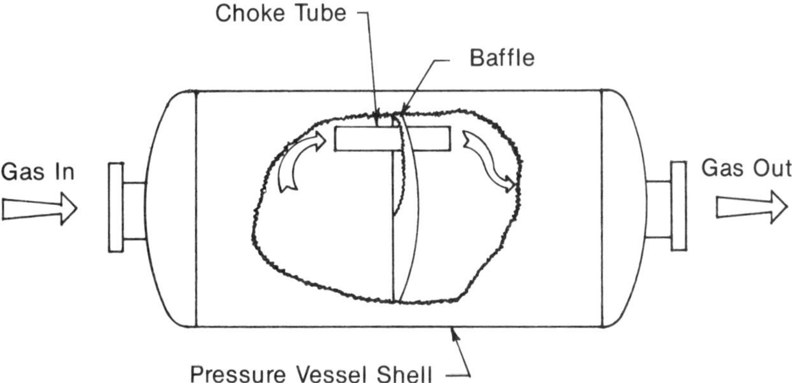


*Figure 52.* **Cutaway of typical pulsation damper**

*Compressors for Refinery Services*, requires that the peak-to-peak pulsation not exceed 2 percent at the terminal connection, but states that lower values should be adopted for high pressures. Rule of thumb prescribes the allowable maximum percentage of residual peak-to-peak pressure pulsation amplitude as:

$$\Delta p_r = \left(\frac{3\ 375}{p_m}\right)^{0.333} \hspace{4cm} \text{Equation 22a}$$

Where:   $\Delta p_r$ = Peak-to-peak pressure pulsation amplitude, percent
           $p_m$ = Mean effective line pressure, lbf/in² absolute

Or:

$$\Delta p_r = \left(\frac{23\ 270}{p_m}\right)^{0.333}$$

When: $p_m$ = Mean effective line pressure, kPa absolute

Thus, the tolerable values for pulsation at 1 000 lbf/in² (6 900 kPa) is 1.5 percent and 3.2 percent at 100 lbf/in² (690 kPa). The shaking force in the piping, however, will be $\Delta p_r$ times the cross-sectional area of the pipe.

## OVERPRESSURE PROTECTION

Each compressor station must have adequate protective devices to ensure that the maximum allowable operating pressure of station piping and equipment, except for American Society of Mechanical Engineers (ASME) pressure vessels, is not exceeded by more than 10 percent. Pressure vessels designed to ASME's Boiler and Pressure Vessel Code (see Appendix A) must be protected according to the appropriate ASME codes. Overpressure protection includes the application of relief valves and shutoff devices. Monitor regulators are used in limited applications, particularly in auxiliary piping systems such as fuel and starting gas.

### Relief Valves

The most common method of providing overpressure protection in a compressor station is by incorporating a relief valve such as the one illustrated in Figure 53, in each segment of gas piping in which an overpressure condition could occur. Since the compressor itself is the primary means of increasing the pressure, the most obvious location

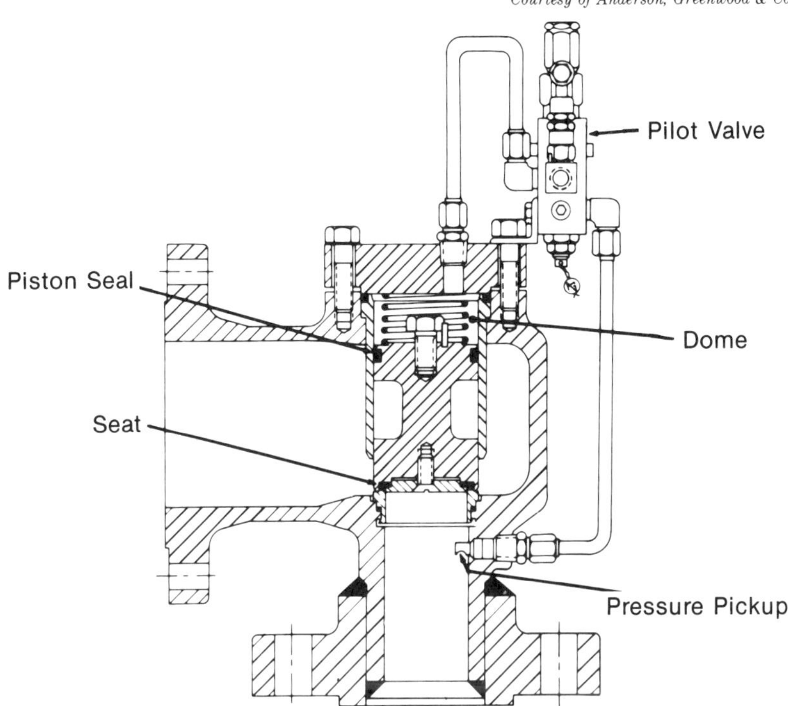





**Figure 53.** Cross-section of a pilot-operated relief valve

for a relief valve is directly downstream of the compressor stage and ahead of the unit discharge valve. Other less obvious locations where relief valves may be used are in the fuel system; in lower pressure suction lines where the discharge gas inadvertently could leak into a bottled-in segment; and in vessels or other segments of piping that could be subjected to excessive pressure from thermal expansion created by ambient temperature fluctuations.

The following design considerations should be evaluated when selecting and sizing a relief valve:

- Maximum flow rate of gas causing the overpressure condition should be determined. Using 110 percent of the nominal design pressure as the full-open position for a pilot-operated relief valve, an adequate relief valve is selected from charts or graphs provided by valve manufacturers.
- If possible, a point near the pressure source should be selected to install inlet piping for the relief valve. This piping must be free

from restrictions that would diminish flow to the relief valve. Inlet piping to the valve must have greater flow capacity than the valve. Otherwise the piping would replace the valve as the critical, restricting pressure-control device.

- It is best if the gas can be exhausted directly into the atmosphere from the relief valve. If it is necessary to flow the gas through a line to vent it in a safe place, care must be taken in sizing the vent line. The vent line must carry the gas from the relief valve to the venting location without creating a back-pressure that will restrict the flow through the relief valve.

### Shutoff Devices

Nearly every compressor is equipped with a series of safety shutdown devices. These devices are used primarily to prevent the equipment from suffering mechanical damage caused by abnormal operating conditions such as high jacket water temperature or low lubricating oil pressure. Generally the safety shutdown system includes high-gas-pressure shutoff devices as well. A high-pressure shutdown control on the discharge gas can prevent most overpressure conditions in a com pressor station long before the relief valve is activated, thus warning the operator of a potentially unsafe condition.

# GAS CLEANING

Practically all natural gas streams carry some contaminants in varying amounts. It is extremely important that careful consideration be given to the selection, installation, and maintenance of adequate gas cleaning facilities at compressor stations to protect the compressors and their ancillary equipment from serious problems that result from operating with dirty gas.

Where reciprocating compression equipment is involved, dirt and entrained liquids can cause valve plate breakage, erode valve seats, and jam valve springs in compressor suction and discharge valves. Contaminants also can accelerate wear of piston rings, ring grooves, and cylinder liners by destroying lubrication. In extreme cases, an entrained liquid can fill the cylinder clearance volume, resulting in serious rod damage, breakage of head studs, or total destruction of the unit.

Centrifugal compressors have a better chance of survival than reciprocating compressors, but dirty gas can cause erosion of the impeller and diaphragms, along with rapid wear of shaft seals. Accumulation of dirt on the impeller can cause reduced pumping capacity and impeller imbalance.

Whatever the type of equipment used – solid, liquid, vapor, or corrosive – gaseous contaminants in the gas stream must be removed to avoid loss of efficiency, excessive downtime, and high maintenance costs. Proper gas cleaning can reduce or eliminate all of these problems.

## TYPES OF GAS CLEANERS

Separators, scrubbers, and filters are the three major types of gas cleaners available to remove contaminants from gas streams. Combinations of these separate types are available which provide improved gas cleaning performance over the individual types. A description of the basic types will help to explain how they work.

### Separators

A separator is a type of gas cleaner that does not use any auxiliary media such as scrubbing oil or filter elements to remove contaminants. Separators are built in many different configurations, but all rely on gravitational, inertial, or centrifugal forces to accomplish separation. Gravitation is utilized in horizontal separators by taking advantage of the downward trajectory of particles in a flowing gas stream. Commonly impingement separators are used which contain baffles that redirect the gas stream a number of times so that the momentum of the particles causes them to strike the baffles and drop out into a reservoir below. Figure 54 illustrates a cross-sectional view of an impingement separator. As gas passes through the vanes, the particles are trapped in the U-shaped baffles and fall into a collection trap below. Centrifugal separators, as the name implies, utilize centrifugal forces to remove entrained matter. They may contain one or many vortex-creating devices which swirl the gas at a velocity great enough to accomplish centrifugal separation. Figure 55 shows a cutaway of a simple, vertical, centrifugal separator with several vortex tubes.

Most separators contain some mix of the above separation methods, but in every case the contaminant must have sufficient mass to be affected by these gravitational and inertial forces. Obviously particulate matter in the low-micron range, liquids in the form of extremely fine mist or vapor, some aerosol mixtures, and corrosive gases will pass through separators with little or no separation taking place.

### Scrubbers

Because of the limitations of separators, gas scrubbers, such as those shown in Figure 56, were devised. In addition to containing some of the elements of separators, they always make use of some added media,

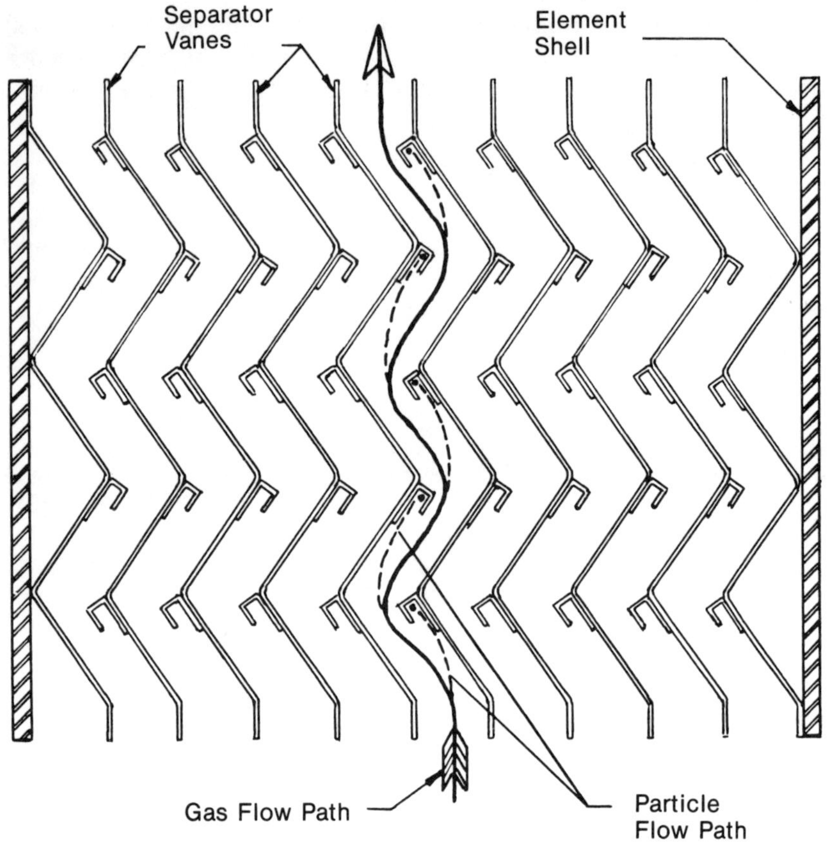


*Figure 54.*    **Cross-section of horizontal impingement-type separator**

usually scrubbing oil, to wet the small particles and suspend them in oil as the gas stream passes through the scrubbing oil. The media containing the contaminants then is removed from the scrubber and cleaned by gravity settlement, filtration, or heating to reduce viscosity and to remove gases. The clean scrubbing media then is returned to the scrubber to complete the cycle. In some scrubbers gas enters the vessel and is intermingled with scrubbing oil by means of one or more vortex-creating elements. Then the gas stream is stripped of entrained scrubbing media by separating elements, which are either of the centrifugal or impingement type. The removed scrubbing media is drained into a reservoir and recycled continuously. Other gas scrubbers may make use

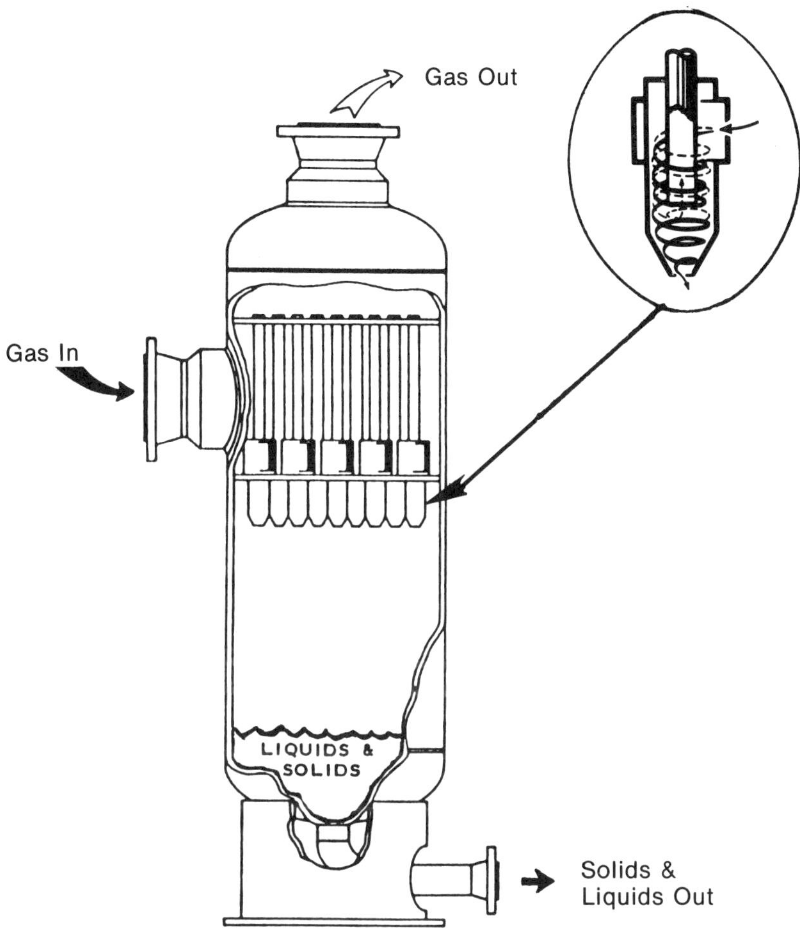



*Figure 55.* **Cutaway of vertical, centrifugal gas separator**

of rotating screens which are dipped into an oil bath. The fine dirt particles cling to the wet screen and are washed into a reservoir as the screen turns. Figure 57 illustrates a filter of this type, in which a vane bundle is rotated slowly by means of an electric motor. Usually a secondary cleaning element is located downstream to catch entrained scrubbing oil. Unfortunately practically all scrubbers permit some carry-over, and while most of the harmful contaminant material has been removed, some small quantities of entrained scrubber oil remain in the gas stream and are carried to equipment downstream.

Courtesy of The East Ohio Gas Company

*Figure 56.* Typical gas scrubbers

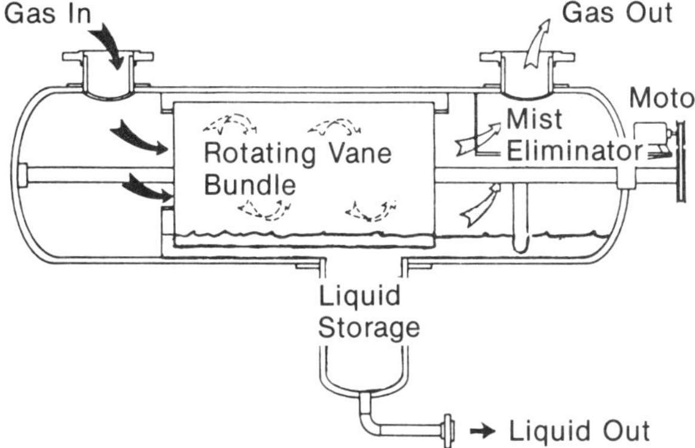


***Figure 57.*** **Cross-section of a motor-driven, rotating vane scrubber**

## Filters

Gas cleaners which make use of filter media, such as the one shown in Figure 58, are available for a wide variety of applications. Generally the filters use natural or man-made fibers in pads or cylindrical elements. Many of these elements are constructed to provide in-depth filtration in which the front portion facing the flow is woven more loosely than those on the exit side. This permits the filter to load more uniformly and thus retain more dirt. Frequently the filter fibers are treated with a tacky material to help retain the dust particles. While dirty filters may actually do a better job of cleaning than brand new filters, as the filter loads, the pressure drop across it increases. Most filter equipment is provided with a gauge or gauges to read the pressure drop across the filter media. Careful observation of this differential pressure is important, and a media change is necessary when it reaches the manufacturer's recommended maximum pressure drop. A reduction in compression efficiency and collapsing of the filter elements may result if these recommendations are not followed.

Some provision for handling the flow during downtime must be made, either in the form of multiple units or temporary bypass piping, since the unit must be taken out of service to replace filters.

Most filter systems incorporate separator elements to remove heavy dirt particles and liquids from the gas stream before they reach the actual filter element. Some units use the filter elements to both clean the gas stream and to agglomerate oil mist into small droplets having enough mass that they can be trapped in an impingement separator element placed downstream from the filter elements. A combination filter-separator is shown in Figure 59.

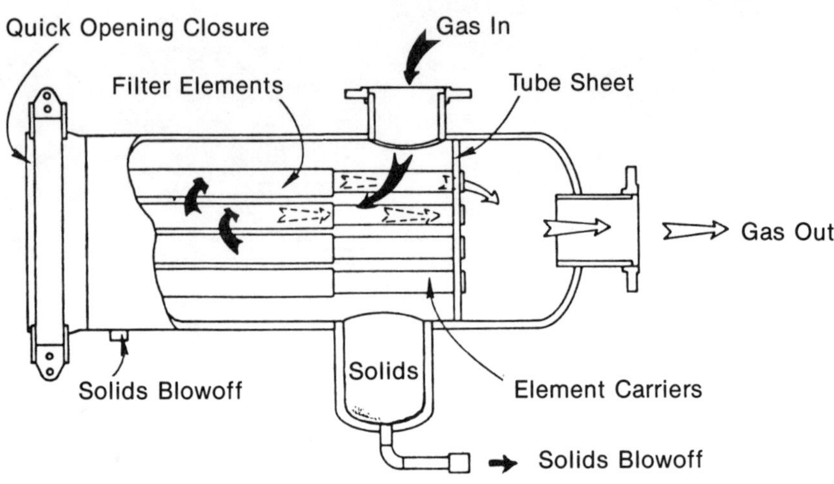


*Figure 58.* **Horizontal gas filter showing cylindrical type filter elements**

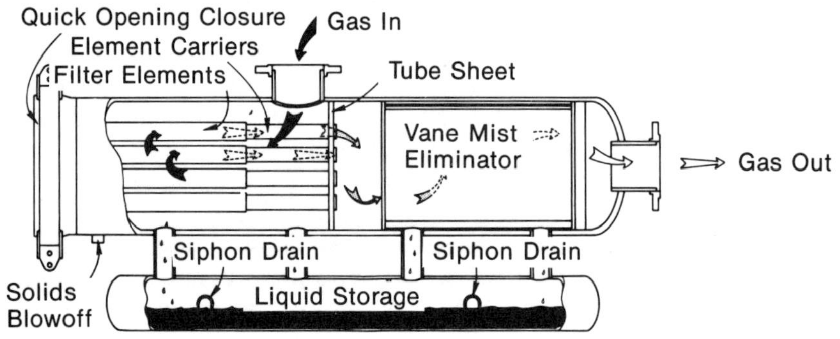


*Figure 59.* **Horizontal filter-separator to remove liquids as well as solids**

## APPLICATION FACTORS FOR GAS CLEANERS

Gas streams can vary from being relatively clean to very dirty. Contaminants include sand, crude oil, natural gasoline, paraffins, hydrocarbon liquids, salt water, rust, and mill scale.

As in many other situations, the problem must be recognized before it can be solved. Many application factors must be considered before a satisfactory solution to any given gas cleaning problem can be reached.

## Gas Cleaner Sizing Criteria

Separators, scrubbers, and filters have optimum flow rates for various pressures which have been determined by the manufacturer after extensive testing. To properly size and select the gas cleaning equipment, the volume of gas to be handled, the maximum working pressure, and the allowable pressure drop must be known.

Frequently it is necessary to install multiple units to permit varying flow conditions and cleaning. Generally the fewer the units, the lower the overall installation cost will be. However the additional cost of several smaller units may be justifiable, especially where a turndown ratio is a valid consideration.

Consideration should be given to the possibility of future plant expansion at the time of original construction. Under these circumstances, providing more capacity than initially is required may turn out to be a good investment. If the possibility for expansion exists, it is prudent to make provisions for future units.

Sizing equipment to operate at a pressure drop of 1 to 2 lbf/in² (7 to 14 kPa) is considered good practice. Using pressure drops higher than 5 lbf/in² (34 kPa) will result in smaller and less costly equipment, but will increase compression costs for the life of the plant and may turn out to be a poor investment.

In summary, generally the specific information needed to size a gas cleaner is as follows:

- Flow rate in MMcf/d (m³/d), Mcf/h (m³/h), or lb/h (kg/h)
- Maximum, normal, and minimum operating pressures
- Relative density of the gas being handled
- Relative density of liquid to be removed
- Amount of liquid to be removed in lb/MMcf (mg/m³), or gal/MMcf (mL/m³)
- Size of particulate matter to be removed
- Amount of particulate matter to be removed in lb/MMcf (mg/m³)
- Operating temperature of the gas stream
- Maximum design temperature and pressure established by the American Society of Mechanical Engineers (ASME) *Code for Unfired Pressure Vessels* (see Appendix A)
- Auxiliary equipment required such as level control dump valves, gauge glasses, valves
- Inlet and outlet nozzle sizes
- Orientation of inlet and outlet nozzles.


## Contaminants

Some of the contaminants found in natural gas streams occur

naturally while others enter the stream because of processing, malfunctions of processing equipment, or overloading of cleaning equipment. It may be in the form of dry particulate matter, liquid, corrosive gas, vapor, aerosol mixture, or combinations of the above. In any case, it is necessary to determine as accurately as possible the nature and quantity of foreign matter which must be removed before a satisfactory selection of gas cleaning equipment can be made.

Dry particulate matter found in gas streams usually consists of silicon, iron oxide in various forms, and dust from dry desiccant beds used in dehydration equipment. Depending on its size, particulate matter 15 micrometres (microns) and larger may be removed by separators. For smaller particles, filters may be used. Depending on the amount of dirt to be removed, consideration should be given to the use of primary separators for bulk removal of foreign matter, followed by filters to trap the smaller particles which escape the primary separators.

Liquids most commonly found in gas streams are crude oil, natural gasoline, and salt water. Usually they can be removed with properly sized and designed impingement and centrifugal type separators. These liquids frequently exist as droplets carried along in the gas stream, but slugging can occur where accumulation of liquids in low points of a pipeline become deep enough to be picked up and moved with the gas flow. Simple drip tanks can be used as slug catchers. However, good separators are needed to catch small droplets. Where fogs or aerosols exist, filters are generally effective. It may be prudent in some cases to provide primary separation equipment to remove most of the entrained liquid before passing the gas through a filter, if large quantities of liquids are present in the gas stream.

Corrosive gases such as hydrogen sulfide, which may occur in natural gas streams, always require special treatment. Generally removal is accomplished by wetting the gas stream with amine solutions or other chemicals in contactor towers, and then separating the liquid from the gas stream with suitable separator elements. Dry chemicals also may be used to remove corrosive gases.

Hydrocarbon vapors are similar in some ways to corrosive gases in that they cannot be removed by ordinary mechanical means. Here again some media, such as oil, must be used to absorb the vapors. This is accomplished by bubbling the gas through absorber oil in contactor towers. Other methods, such as stage separation or low-temperature separation, may be used. These methods always involve sizable pressure reductions to accomplish separation, and are economical only under specific operating conditions.

Aerosol mixtures can be handled best by filters because of the coalescing action which takes place within the filter media. These mixtures can prove to be a very difficult problem, and recommendations from reputable gas cleaning equipment suppliers should always be sought in these cases.

## Gas Cleaner Operating Characteristics

The operating characteristics of different types of gas cleaners are dependent on a number of factors such as location, piping, gas flow, temperature, pressure, and maintenance, to name the more important ones. A brief discussion of these factors follows.

Usually a gas cleaner is located upstream of the equipment it must protect. At a compressor station it would be placed on the suction side of the compressors. In the case where gas measurement equipment is to be installed on the suction side of a reciprocating compressor station, gas cleaners located between the orifice meters and the compressors could provide some pulsation damping to improve measurement. However, the gas passing through the meters would be dirty, and fouling of the orifice plate could result in less overall accuracy of measurement than would be obtained with the measurement downstream.

Sometimes it may be necessary to provide gas cleaners downstream from compressor equipment. For example, where dehydration equipment is installed on the discharge side of a compressor unit, lube oil from the compressors could foul the desiccant used in the dehydration equipment. Location of a gas cleaner between the compressor and the dehydration equipment is indicated under these circumstances. Another important consideration is to locate gas cleaners where they readily can be maintained, and where ample provisions can be made for storing the removed contaminants.

Gas flow through a cleaner can affect its operating characteristics. Pulsation from a reciprocating compressor located downstream may result in a loss in the cleaning efficiency. If the gas velocity is too high or too low, the performance of the gas cleaner may be affected adversely.

In most cases gas entering a cleaner will be near the soil temperature, which is about 50°F (10°C). High ambient temperatures rarely present a problem. Low temperatures, however, can affect the operating characteristics of a gas cleaner. In scrubbers, oils with low pour points can prevent some problems. In liquid separators crude oil with high pour points can be difficult to remove.

The oil in scrubbers must be monitored and maintained. Too much oil in a scrubber will increase carry-over, and too little oil will cause poor cleaning efficiency.

The performance of gas cleaners is related directly to the maintenance provided. The time and cost of taking care of this equipment will be repaid many times in the successful operation of the downstream compressor station and other equipment.

## DUMP SYSTEMS

Foreign matter separated from the gas stream accumulates in the gas cleaner and has to be removed periodically. Dump systems perform this important function. While there are many different systems from which to choose, most of them contain similar elements such as dump valves, level controllers, inter-piping, and storage reservoirs. Careful consideration in the design of these elements is essential for a satisfactory dump system.

# GAS COOLING

Gas cooling equipment is necessary in compressor stations to remove heat from the gas stream between stages of compression or after the final stage of compression. Heat is transferred from the hot gas to some other cooler medium, usually air. Some of the different configurations of heat exchangers used are described below.

## HEAT EXCHANGERS

The most common heat exchangers found in compressor stations are those used to remove heat from gas streams, reduce cooling water temperature, lower oil temperature, or to cool compressed air. When a heat exchanger is installed in a gas or air stream between stages of compression, it is referred to as an intercooler. A heat exchanger located in downstream piping of a compressor is called an aftercooler. Both lower the stream temperatures but for entirely different reasons. Intercoolers reduce the temperature rise caused by the heat of compression. The discharge temperature for a particular stage of compression can be calculated using Equation 12. The results of this computation are ideal and generally somewhat low because the formula neglects friction, irreversibility, and other factors.

If the heat of compression is not reduced by an intercooler, the second stage compressor will operate at a lower efficiency because of the expanded hot gas it must handle. At high ratios of compression without intercooling, the excessive heat can destroy compressor lubrication, damage compressor valves, and ruin rod packing in reciprocating compressors. It is good practice to limit the compression ratio per stage to four, and intercooling should be provided.

Aftercoolers are needed for other reasons, primarily to preserve the integrity of the protective coating commonly used on the discharge piping. Because of its low melting point, exposure to high temperatures will cause the coating to sag, thereby reducing its effectiveness. Special coating for higher temperatures may be used, but in general it is good practice to limit discharge temperatures to 125°F (52°C).

Decreasing the gas temperature with an aftercooler also reduces high stress levels in discharge piping created by thermal expansion. Temperature-induced stresses can cause major problems at bends and branch connections if sufficient aftercooling is not provided, and stress corrosion can be accelerated by elevated temperatures.

Gas coolers are an essential part of most compressor stations. A description of the various types will help clarify their selection and use.

## Types of Gas Coolers

Several types of gas coolers are available, but the aerial cooler is one of the most common currently found in compressor stations. Some shell-and-tube gas coolers or exposed tube sections sprayed with cold water are also in use where high ambient temperatures make aerial coolers impractical. Because of the completely different configuration of these coolers, they will be discussed separately.

Aerial coolers, as their name implies, make use of air to cool hot gas. There are two kinds of aerial coolers, namely forced draft and induced draft, as illustrated in Figure 60. Figure 48 shows a typical compressor station layout and how the gas cooler is situated physically in the gas piping arrangement.

In compressor stations forced-draft aerial coolers are used more frequently than induced-draft for a number of reasons. The larger coolers usually consist of one or more tube bundles horizontally supported by structural members, which also support a plenum chamber. Each tube bundle consists of two headers, as shown in Figure 60, spaced by structural members. The headers are connected by a number of finned tubes which transport the hot gas from one header to the other. Arranged in horizontal rows, the extended surface of the finned tubes permits them to extract heat from the gas, and transfer it to the air stream which is forced over them by a fan located below. To minimize vibration the fan and its drive sometimes are supported on a separate foundation independent from the main cooler support.

The tube bundles are an extremely important part of the cooler. Operating pressure dictates the design of the headers and tubes. Carbon steel tubes are usually satisfactory for most applications and are required for high-pressure service. Aluminum fins are attached to the

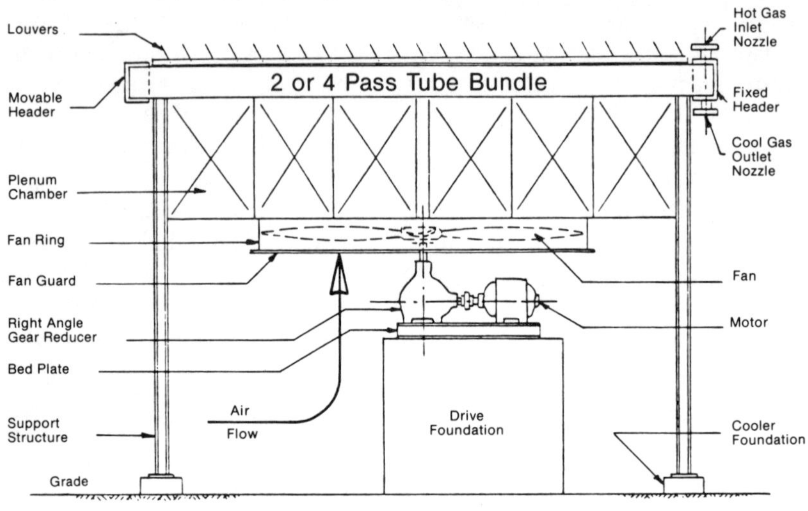


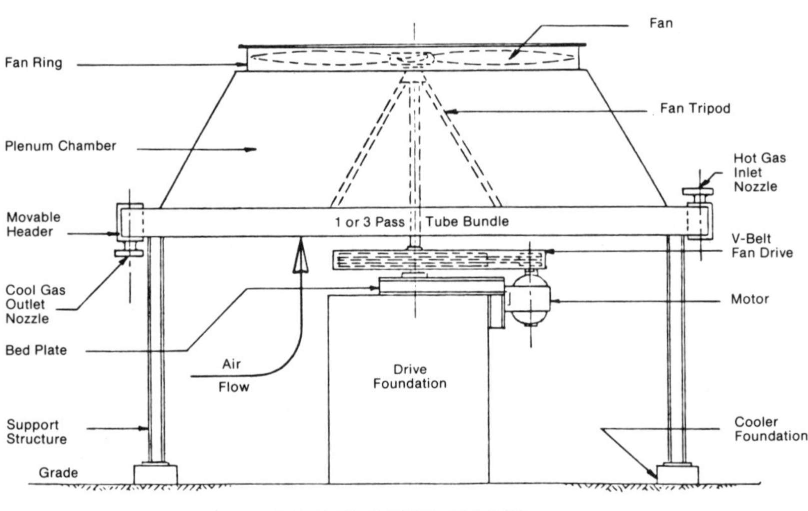


*Figure 60.* Types of aerial heat exchangers

tubes to increase the heat transfer surface. Preferably the number of rows of tubes in the bundles should not exceed three since the air stream is heated by the rows of tubes facing the fan, and the temperature differential becomes less at each successive row. Cleaning the exterior surface of the tubes also becomes more difficult as the number of rows increases.

Induced-draft aerial coolers basically have the same components as the forced draft, but they are arranged so that the fan is mounted above a horizontal tube bundle; and the ambient air is pulled rather than pushed across the finned tubes, resulting in a better air distribution. Recirculation is less likely to occur with induced-draft coolers, and the noise level is somewhat less. With the plenum chamber and fan mounted above, the tube bundle is protected from rain and hail that can affect cooler performance and damage fins. Both induced-and-forced draft aerial coolers will perform satisfactorily, but the forced-draft type seems to be preferred for compressor station installations.

Shell-and-tube gas coolers have a limited application in compressor stations. Heat is removed from the gas stream to the water, which then is cooled in a cooling tower and recirculated. This is a more expensive way to cool gas but is practical when high ambient temperatures make it impossible to cool the gas sufficiently with air.

## Gas Cooler Sizing Criteria

Sizing gas coolers is a difficult job because of the many variables that exist at a given installation. Two important items that must be determined are the actual heat load and the design air temperature. The heat load may be found from the following:

$$Q_\mathrm{H} = q \times \rho \times c_p \times (t_1 - t_2) \qquad \text{Equation 23}$$

Where:  $Q_\mathrm{H}$ = Heat load in desired units, Btu/h or W
        $q$ = Gas flow, scf/h [m³/s (st)]
        $c_p$ = Specific heat capacity of gas at the average of the inlet and outlet temperatures, Btu/(lb•°F) [J/(kg•°C)]
        $t_1$ = Gas inlet temperature, °F (°C)
        $t_2$ = Gas outlet temperature, °F (°C)
        $\rho$ = 0.0765×$d$, lb/scf, or 1.2247×$d$, kg/m³(st), and
        $d$ = Relative density

The design air temperature is usually the average daily maximum temperature for the hottest month of the year at the gas cooler location. Because some recirculation of air may occur, it is good practice to add from 3 to 5°F (1.5 to 2.5°C) to the average temperature depending upon the location of the cooler with respect to other coolers, buildings, trees, and other environmental conditions.

Additional information necessary for cooler design is as follows:
- Type of gas cooler (forced-draft or induced-draft)
- Normal operating pressure
- Maximum design pressure
- Maximum allowable pressure drop
- Fouling factor (tube and air sides)
- Tube and fin material
- Header material and type
- Type of plugs desired
- Number of passes desired
- Size and pressure rating of header nozzles
- Construction code required for headers
- Test pressure for header
- Type of fan (fixed or variable pitch)
- Fan blade material
- Type of fan drive
- Noise level requirements
- Accessory equipment.

The approach factor, the difference between the outlet gas temperature and the air temperature, is an important consideration also. The lower the value specified, the more surface will be required in the tube bundle, and the larger and more costly it will become. Usually approaches less than 15°F (8.5°C) are economically impractical.

Since the heat given up by the gas is equal to the heat gain of the air, Equation 23 may be used to calculate the required air mass flow rate needed to select the gas cooler.

## TEMPERATURE CONTROL

Some gas cooler installations do not require temperature control, and the lower the gas outlet temperature, the better. There are other situations, however, where temperature must be controlled to conserve energy, or for specific process reasons. The most common methods of temperature control for both forced and induced-draft coolers involve the use of adjustable louvers, variable fan speed control, variable fan blade pitch control, or combinations of these methods.

### Adjustable Louvers

Louvers mounted on the face of the tube bundle provide temperature control by restricting the flow of air through the tube bundle. These louvers may be positioned from fully closed to fully open, either manually or automatically. Automatic control is accomplished by using hydraulic, pneumatic, or electric motor operators. A temperature sensor located in the gas piping downstream from the cooler is connected to a temperature controller, which in turn signals the louver operator to position the louvers so the proper temperature is maintained. In addition to controlling temperature, these louvers provide hailstorm protection if they are installed on top of the tube bundle and, if properly directed, can help reduce recirculation.

## Variable Fan Speed

Fan speed control is a practical means of providing temperature control. If electric motors are used and precise control is not necessary, two-speed motors should be considered. Most motors require only one-quarter of the power consumption at half speed as compared to full speed. Some electric motors with sophisticated electronic control equipment are capable of providing infinitely variable speeds. Variable speed is also possible with hydraulic fan drives or with fans driven by gas, gasoline, or diesel engines. Turbine drives can provide variable fan speeds also.

## Variable Fan Blade Pitch

Varying the fan blade pitch is another method used to change the air flow through the tube bundle and thereby control the gas temperature. These devices are usually hydraulically or pneumatically operated, and although they conserve power, they are relatively expensive to buy and maintain. Their use should be considered only when accurate temperature control is essential.

Combinations of the three methods are suggested for certain installations. The initial investment, as well as the operating and maintenance costs, should be considered when selecting a method of temperature control for any given installation.

## FAN DEVICES

Fans and their associated drive mechanisms are a vital part of any aerial gas cooler. The fan diameter can range from 3 to 28 feet (900 to 8 500 millimetres), but about 15 feet (4 600 millimetres) is the maximum diameter to consider for ordinary design conditions. As a rule a tip speed of 10 000 ft/min (51 m/s) should not be exceeded because of the noise inherent in the fan operation. For locations where noise can create environmental problems, 9 000 ft/min (46 m/s) tip speed should be the maximum. As an additional consideration, induced-draft fans will produce less noise than forced-draft fans.

Fan blade materials may be aluminum, laminated or molded plastic, carbon steel, stainless steel, or monel metal. Generally plastic blades are not used in induced-draft coolers because of the higher temperature of the air being moved. However, plastic blades tend to operate more quietly than metal blades and are used in some forced-draft installations. Aluminum blades frequently are used where atmospheric conditions are not corrosive.

Because the fan speed usually is slower than its driver, some form of speed reducer must be used. Perhaps the least expensive method is a V-belt drive, but it should not be considered for installations over 20 hp (15 kW), or where fan diameters exceed 10 feet (3 050 millimetres). Right-angle gear reducers commonly are used in fan drives because of their greater horsepower capacity and because they allow the driver to be mounted horizontally.

Occasionally tripods are required to support the fan at some distance above the drive to provide clearance between the drive and the plenum chamber. Tripods should be designed carefully to minimize vibration and to support the fan concentrically in the fan ring. Excessive fan ring clearance can diminish the performance of the fan. Depth of the fan ring is also important, and the vertical location of the fan in its ring must be maintained as specified by the supplier to ensure proper operation. Figure 61 illustrates four of the most common types of fan drive arrangements used in compressor station gas cooler applications.

**NOTE:** *Fans and their associated drive mechanisms should be designed and located to facilitate physical inspection at regular maintenance intervals.*

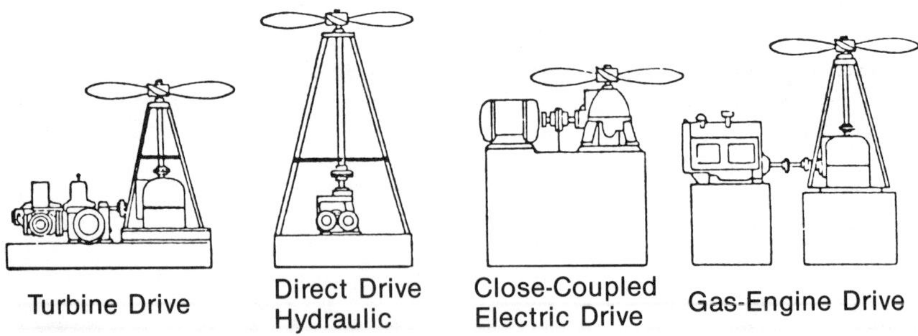


*Figure 61.* **Examples of cooler fan drive arrangements**

# COMPRESSED AIR FACILITIES

Compressed air systems consist of four major components: compressors, tanks or receivers, piping, and conditioning equipment. For proper design it is necessary to consider the required capacity, discharge pressure, and type of compressor for each application.

## AIR VOLUME REQUIREMENTS

The air volume requirements of the system must be balanced between engine starting, utility, and instrument air needs. Engine starting consumes the largest quantities of air, followed by utility air needs. However, the operation of control valves and instrumentation is vital to the operation of the station, mandating dedication of air for instrument control with a system recovery time of between 15 and 30 minutes. This is a major factor in determining the size of an air receiver since the air compressor output is small compared to the demand for air during the first 5 to 30 seconds of an engine start.

Future plans for the station should be considered, and if expansion is expected within 2 or 3 years, it may be advisable to install an oversized system initially rather than adding to the system at a later date.

Although it is the responsibility of operators to eliminate leakage, the fact remains that air losses will occur because of leaks in joints, tools, and hose connnections. Therefore, an allowance of approximately 10 percent of the total air requirements should be applied to account for losses caused by leakage and deterioration.

The selection of a compressed air system involves an economic analysis of operating and maintenance costs, energy consumption, and installed system costs. Many of the design factors interrelate with each other. Considering that the consumption of compressed air constitutes a significant portion of the auxiliary energy cost, it is imperative that each component be designed specifically for a given application.

### Engine Starting Air

The size of a compressed air system is primarily dependent upon the requirements for compressed air consumed in engine starting. Considering the number and size of engines installed, volumetric rate of air, and pressure requirements for each engine, the capacity of the compressed air system can be determined.

The starting air system usually consists of two half-size, light-duty, air-cooled compressors with appropriately sized receivers. The total air compressor capacity should be sized to bring the receivers from 120 to 250 lbf/in² (830 to 1 720 kPa) in 15 minutes. In this pressure range the driving power requirement is about 27.5 hp per 100 ft³/min (43.5 kW per 100 L/s) airflow. The time interval that an air receiver can supply air from a certain initial pressure to a specific final pressure at a constant flow rate may be calculated by using the following equation:

$$t = \frac{V \times (p_i - p_f)}{Q \times p_{amb}} \qquad\qquad \text{Equation 24}$$

Where:   $t =$ Time in desired units, min or s
$V =$ Volume of air receiver, ft³ (L)
$p_i =$ Initial pressure, lbf/in² (kPa)
$p_f =$ Final pressure, lbf/in² (kPa)
$Q =$ Airflow rate, ft³/min, (L/s)
$p_{amb} =$ Atmospheric pressure, lbf/in² (kPa) absolute

Two light-duty compressor units provide the required redundancy, and their cost is lower than that of a single heavy-duty unit. Standard drive is electric with automatic start/stop control; however, one drive may be a reciprocating engine if electrical service continuity is of concern. The installation of two receiver tanks is a good practice in any installation. For good design the total receiver capacity is sized to allow for three start attempts every 20 minutes for the largest unit in the station. The starting air requirements may be estimated according to the following information:
- Naturally aspirated engines–10 ft³ per 100 bhp (0.38 L/kW)
- 4-stroke turbocharged engines–15 ft³ per 100 bhp (0.56 L/kW)
- 2-stroke turbocharged engines–25 ft³ per 100 bhp (0.95 L/kW)

### Utility and Instrumentation Air

Utility and instrument air requirements constitute the bulk of daily air usage. All air-operated equipment, such as pneumatic control valves, air motors, pumps, tools, etc., should be considered in designing the utility air system. Rates of air consumption can be obtained from the equipment manufacturer. Air consumption factors dealing with load percentages and air demand can be found in compressed air and gas industry handbooks.

## AIR COMPRESSORS

Reciprocating air compressor units operating at 275 lbf/in² (1 900 kPa) and producing from 32 to 8 475 ft³/min (15 to 4000 L/s) of air are used most commonly in compressor station installations. These units can be single- or multiple-staged with air or water cooling. They can be specified with either lubricated or non-lubricated pistons and cylinders. A non-lubricated system with teflon rings and riders will provide an oil-free air supply. Because the risk of an air line explosion is increased by lubricating oil, a synthetic oil with a high flash point is recommended. A compatible trim will be necessary to prevent deterioration of rings and gaskets.

Dependability, cost, efficiency, and availability of energy source are factors to be considered in the selection of air compressors. Electricity rates, time of day, demand peaks, power outages, and power factors may influence the type of drive and hours of operation.

Multiple compressor units increase the dependability of a continuous air supply over single-unit systems. This increases capital investment, however. Where multiple units are used, they should be designed to operate in sequence to ensure uniform operation. In addition to the main units, a small, gas-driven compressor sized to maintain the minimum air requirements of the station should be included for emergency standby service.

## Air Compressor Operating Modes

Where large volumes of air are required, usually it is more economical to operate a compressor continuously rather than in a frequent on/off cycle. Therefore, an air compressor with a variable discharge rate capability should be considered.

In situations where a significant amount of instrument air is required, a smaller, non-lube compressor capable of 75 to 100 lbf/in² (520 to 690 kPa) discharge pressure is utilized in the continuous-duty, variable-discharge, unloaded operation described above. A standby supply for instruments is provided from the higher starting air system with a regulator set to feed at pressures less than that normally experienced when the instrument air compressor is functioning properly.

## Air Storage and Piping

The amount of air storage required should be evaluated from an energy viewpoint keeping in mind that the amount of energy available is dependent on the air pressure, and that starting the second and subsequent engines may be utilizing air at diminishing storage pressures and volumes.

The operating pressure differences between the various air uses require a well-maintained system of regulators and check valves. Compressed air piping should be designed according to the American National Standards Institute (ANSI) Code B31.1, *Power Piping,* Chapter II, Part 1, for pressure piping (see Appendix A).

Engine starting lines should be sized to provide large air volumes without a significant pressure loss. An NPS 3 or NPS 4 Schedule 40 steel line is usually sufficient. Utility lines are reduced to NPS 2 operating at 100 lbf/in$^2$ (690 kPa). This requires a pressure regulator and excess flow safety shutoff valves. The drop lines may be NPS 1 steel pipe and will usually have quick-disconnect couplers and shutoff valves. A back-pressure regulator should be included between the primary air receiver and the auxiliary receivers to maintain a minimum station air supply of 100 lbf/in$^2$ (690 kPa) for control air. Piping for pneumatic controls and instrumentation should be relatively small, for example: 0.25-inch (6 millimetre) I.D. at low pressure [3 to 22 lbf/in$^2$ (20 to 150 kPa)] and at low volumes. The use of brass, copper, or stainless steel line is dependent upon the application. Pressure drops across the headers, drop lines, and check valves must be evaluated for their effect on the total pressure requirement of the system.

## CONDITIONING OF COMPRESSED AIR

To operate properly, pneumatic controls and instrumentation require dry, oil-free air. They also must be free of any foreign matter greater than 10 micrometres (microns) in size. A clean air system should have an intake filter for dust removal, preferably located outside the building; a storage tank; an aftercooler or chemical drier; and water traps in the lines for moisture removal. Filters to remove oil from the air also should be considered. Unless these factors are accounted for, equipment maintenance costs may increase, as well as the risk of an oil-related explosion. Non-lube compressors furnish air with fewer contaminants and less risk of explosion in the system.

## AIR RECEIVERS

A proper balance between the air receiver and the compressed air usage can optimize installation, operation, and maintenance costs. The receivers must be designed in accordance with the American Society of Mechanical Engineers (ASME) Section VIII, *Boiler and Pressure Vessel Code* (see Appendix A). They should be equipped with a manway and drains, as well as shutoff, relief, and bypass valves. Typical air storage tanks are shown in Figure 62.

**Figure 62.** **Vertical air storage tanks**

# ELECTRIC POWER AND LIGHTING

Successful compressor station operation depends heavily on the station electrical system. The following addresses engineering and design considerations for compressor station electrical systems.

Historically electrical systems for mainline compressor station were designed with two criteria in mind. One was to provide the most reliable form of purchased power available. Another consideration was to provide a source of on-site power generation to operate independently of the electric utility. Through the years these two criteria have been met in many different ways, and are still applicable to the design of compressor station electrical systems today.

Personal safety and protection of property from electric arcing, fire, and explosions which may result from the installation and use of electric equipment should be considered carefully during design. State and municipal regulations vary, and fire and casualty insurance carriers usually have definite requirements of their own. Details of the proposed installation should be reviewed with representatives of these various groups before construction is started.

In general all electrical construction should conform to the latest requirements of the *National Electrical Code* and the *National Electrical Safety Code*. Both of these publications are American standards and approved by ANSI. (In Canada the counterpart of NEC is the *Cana-*

*dian Electrical Code* published by the Canadian Standards Association.) The National Electrical Code, ANSI/NFPA 70, has been incorporated by reference into many regulations of construction promulgated by the Department of Labor. While the word "approved" is defined as "acceptable to the authority having jurisdiction" in the NEC, some federal regulations define it as "of a type approved by the Underwriters Laboratories, Inc., or Factory Mutual Engineering Corporation" for the specific application.

## PURCHASED POWER

Purchased power usually is delivered to the compressor station auxiliary building. Most auxiliary buildings are centrally located and provide a good location to accept service from the electric utility. The auxiliary building also provides a non-hazardous location for electrical switchgear, generators, and other electric motor-driven equipment.

### Transformer Substation

When a compressor station is supplied by a high voltage (above 15 kV) feed, the utility will transform the power to the plant distribution level desired by the customer. The required transformer will be owned and maintained by the utility, but the customer is expected to provide the outdoor space for the required substation.

When the purchased power is brought in at voltages below 15 kV, the substation may be owned and maintained by either the electric utility or the company purchasing the power. The service entrance normally is the demarcation line where the electric utility's responsibility ends and the owner's responsibility begins.

Transformer substations are provided in many configurations. The simplest would be a single transformer mounted on a pole. A more complex arrangement would be a group of transformers, as shown in Figure 63, interconnected in such a way as to provide the precise voltage and the required capacity. Transformer capacity is expressed in kilovolt-amperes. This capacity rating is temperature dependent, and the same transformer may be used at higher ratings when equipped with forced cooling.

Most substations also contain auxiliary equipment such as disconnects, circuit protection equipment, metering, and provisions for grounding.

### Service Entrance

Depending upon the agreement between the electric utility and the

*Figure 63.* **Electrical transformer substation**

owner, the service entrance may be either on the line side or the load side of the transformer substation. In simplest form, it could be a splice box used to connect the power cable from the electric utility to the service entrance furnished by the owner of the facility. Service entrance equipment, however, includes all of the elements needed to protect the source of supply such as disconnects, overcurrent protection, directional protection, grounding, etc. It also includes the metering equipment, certain components of which may be furnished by the electric utility for installation by the owner.

Service entrance design must cover in detail all facilities that are required to carry electrical energy from the electric utility to the distribution switchgear. Facilities may be overhead or underground, and may range from parallel runs of conduit containing several conductors per phase to underground cable or enclosed busways. All service characteristics, such as voltage, frequency, phase, and types of conductors and conductor supports, must be defined clearly. Design of the service entrance for any compressor station may be reduced to a procedure involving five basic steps.

1. Calculate the required current rating and power capacity of the service feeder, once the load analysis is complete.

2. Lay out elements of service connection to the source of the electric utility power.
3. Provide disconnect means and effective overcurrent protection for the service feeders and switchgear.
4. Consider any emergency supply source required for fire pumps, emergency lighting, etc.
5. Consult with the electric utility company, and carefully observe all of the detailed regulations set out in their published standards.

## Distribution Switchgear

Most auxiliary buildings are centrally located, and thus are in an excellent position to serve as electric power distribution centers. The distribution switchgear may serve individual pieces of equipment as well as other load centers. Switchgear design practices logically separate switching and protective devices into medium-voltage systems and low-voltage systems. Recent trends have been toward fewer voltage classes with wider steps between them. Medium-voltage systems are defined as those that operate at voltages greater than 600 but less than 15 000. Low-voltage systems are defined as those that operate at 600 volts or less.

Generally distribution switchgear is located in the auxiliary building of the compressor station. Typically at this location the electrical system changes electrical characteristics or distribution characteristics in order to meet the requirements of electrical equipment at the station.

### Switchgear Structures

Switchgear structures are designed and built to meet industry standards, such as: American National Standards Institute (ANSI), Institute of Electrical and Electronics Engineers (IEEE), and National Electrical Manufacturers Association (NEMA). In order for switchgear to have Underwriters Laboratory (UL) label, components that are used must be UL-listed, and the equipment itself must have been tested as a unit.

Switchgear may be classified as front-connected or rear-connected. Location and accessibility of the switchgear will determine which design will be used. Rear-connected boards are more common at compressor stations today.

Many older compressor station installations have front- and rear-connected gear in which the front gear is live. Live front gear has current-carrying metal parts that are exposed to a person on the operating side of the switchgear. All switchgear designed in recent years is of dead front construction. Dead front construction is defined as being without exposed current-carrying metal parts.

### Switchgear Bus Arrangements

Switchgear bus arrangements must be considered because the bus arrangement determines how the switchgear is designed to operate in conjunction with the equipment being served. There are three common bus arrangements in use today with many variations of each. These are dual-, single-, and split-bus systems.

Dual-bus design involves the use of two independent, identical buses, one for purchased power and one for power being supplied by on-site generators.

Single-bus design involves the use of one bus that is switched from purchased power to the on-site generators by the use of a transfer switch. The bus is energized by one source at a time. Single-bus systems also are used at compressor stations that have no on-site generation and depend upon purchased power as the only source. They are also used on-site where generators are the only source of electricity.

The split-bus design involves the use of a single bus that is split in the middle for insertion of a tie circuit breaker. A split-bus system is used for special applications that would not be feasible with a single-bus or dual-bus design. For instance, one-half of the bus may be reserved to supply critical loads with an on-site generator upon loss of the utility supply (load-shedding).

### Switchgear Devices

Switchgear designed for compressor stations use a combination of individual and group-mounted devices which include power circuit breakers; hybrid, insulated case, power circuit breakers; molded case breakers; fusible switches; and even motor starters. Large equipment may be of fixed or drawout design. Accessories, including switches, meters, indication lights, relays, and ground detection equipment, may be provided and mounted on the gear.

## POWER DISTRIBUTION

In any electrical system the distribution system consists of the facilities used to carry power from the service equipment to the over-current devices protecting the branch circuits. The distribution system for a compressor station may operate at a single voltage level, or may involve one or more transformations.

Design of the distribution system is a matter of selecting circuit layouts and equipment necessary for the conditions of voltage, current, and frequency. The factors of capacity, accessibility, flexibility, and safety must be considered.

Most compressor station electrical systems use simple radial distribution, which receives power from the electric utility at a single substation and steps the voltage down to the utilization voltage. Then feeders are connected to the switchgear through circuit breakers or other protective devices. Figure 64 shows a typical distribution switchgear.

Conventional, simple radial systems distribute power at the primary voltage. The voltage then is stepped down to utilization voltages in several load areas within the station. When power is distributed to the load areas at a primary voltage, energy losses are reduced and feeder circuit costs are reduced substantially.

Courtesy of The East Ohio Gas Company

*Figure 64.* Compressor station electrical switchgear

## Motor Control Centers

A motor control center is an assembly of one or more enclosed vertical sections having a common horizontal power bus and principally containing combination motor control units. Figure 65 illustrates one type of motor control center.



*Courtesy of Allen-Bradley Company*

*Figure 65.* Motor control center

A combination motor control unit includes externally operable, circuit-disconnecting means, branch circuit overcurrent protection, and a magnetic motor controller with associated auxiliary devices. The disconnecting means and branch circuit protection consist of a fusible disconnect or circuit breaker. The motor controller must include motor and branch circuit overload protection unless equivalent protection is provided.

In addition to combination starters, motor control centers may contain lighting panels, feeder tap units, incoming line fusible switches or circuit breakers, lighting transformers, and special equipment assemblies. Motor control centers may be subdivided into Class I and Class II centers.

Class I motor control centers consist of a mechanical grouping of combination motor controllers, feeder-circuit breakers, and other units arranged as a convenient assembly. They include connections from the common horizontal power bus to the motor controllers. They do not include interwiring between units of the motor control center or interlocking with other devices.

Class II motor control centers consist of a grouping of combination motor controls, feeder circuit breakers, and other units designed to form a complete control system. They include all necessary interwiring between units of the motor control center and interlocking provisions to other devices, in addition to connections from the common power bus to the units.

## ON-SITE GENERATION

Most compressor stations have some form of on-site power generating capability. Engine generator sets or gas turbine generator sets are used successfully to provide both primary and standby power.

Selection considerations depend upon whether the unit is to provide primary or standby power. A unit that is to supply primary power should be reliable in continuous service, and have low operating and maintenance costs. The unit should have sufficient capacity to supply power to the connected electrical load needed to operate the compressor station under the maximum load conditions.

Units that are to be used for standby power should have the ability to start and accept loads within 10 seconds after reaching the rated speed, and should have sufficient capacity to supply power for critical loads to operate the station until purchased power is restored. A standby unit can be sized to accept the full station load when desired. Figure 66 shows a standby generator driven by a reciprocating gas engine.

Since 1981 the National Electrical Code (NEC) has addressed what

*Figure 66.* Standby electrical generator

it defines as optional standby systems. Standby power generation at compressor stations falls into this category, and Article 702 of the NEC (see Appendix A) should be followed for these installations.

Gas engine generator sets are available with capacities of from a few hundred watts to several hundred kilowatts.

The advantages of gas-engine generator sets are ease of installation, reliable service, and they are less expensive to operate than simple-cycle, gas turbine generator sets. The disadvantages of the engine generator sets are that they require large foundations, vibration damping, and some form of cooling water system.

Gas turbine generator sets have been used as sources of on-site generation for heavy loads of 670 hp (500 kW) and larger. Their chief advantages include the small size, light weight, low vibration, and little or no cooling requirements. The disadvantages of gas turbine generator sets are that they require provision for large exhausts, are slow starting, may require soundproofing, and generally have high operating and maintenance costs. In addition, it may not be practical to repair gas turbines on-site.

Generator sets should have sufficient capacity to satisfy the maximum compressor station electrical load conditions after some load factor has been applied. They should have the capacity to allow for motor starting, high in-rush loads—as well as future additional loads.

In general it can be said the simpler the control the greater the reliability because there are fewer devices to malfunction or interact with one another. However, certain applications require complex controls for safety and equipment protection, and such controls should be used if the application requires them.

## HAZARDOUS LOCATIONS

A hazardous (or classified) location is one in which the electrical construction practices are prescribed by Articles 500 through 503 of the National Electrical Code (NEC). The code divides hazardous locations into three classes depending upon the kind of hazardous material involved, and divides each class into two divisions according to the degree or severity of the hazard. Also, for testing and approval purposes, various atmospheric mixtures of gases, volatile-liquid vapors, and hazardous dusts have been grouped on the basis of their hazardous characteristics.

Class I locations are those in which flammable gases or vapors are or may be present in the air in quantities sufficient to produce explosive or ignitible mixtures; Class II explosive mixtures will involve dusts; and Class III, fibers or flyings.

Division 1 locations are those in which ignitible concentrations can exist under normal operating conditions, or where faulty operation of equipment or process resulting in ignitible concentrations might also cause simultaneous failure of electrical equipment. Division 2 locations are those in which ignitible concentrations may exist, but only in a distinctly abnormal operating condition.

Group D designation is of primary interest here, as it covers atmospheres containing natural gas. Of minor interest is the Group B designation, as it deals with gases of equivalent hazard to hydrogen such as manufactured gas.

A.G.A. Publication No. XF0277, *Classification of Gas Utility Areas for Electrical Installations*,* is an excellent guideline developed by member companies to aid in classifying hazardous gas utility locations. It is not a code of rules or regulations, but rather a general reference

---

*American Gas Association Operating Section Compressor Committee, *Classification of Gas Utility Areas for Electrical Installations*, 4th ed. (Arlington, 1973), [Catalog no. XF0277].

that may be used as a starting point in determining area classifications in compressor stations. This publication should not be a substitute for good engineering judgment as a means of complying with the National Electrical Code.

## Electrical Sources of Ignition

Equipment such as switches, circuit breakers, and motor starters, containing contacts are sources of ignition for a combustible mixture of gases. Other potential electrical hazards are devices that produce heat, such as lighting fixtures, heaters, or electric motors. Improper design of these devices may result in exceeding the safe surface temperatures for hazardous locations. Also, other parts of the electrical system, such as wiring, transformers, solenoids, and other coil-type devices, could fail and become potential sources of ignition. Therefore, electrical safety is of crucial importance to prevent accidental ignition of an explosive mixture. Compressor station electrical facilities must be designed with this in mind.

## Explosion-Proof Equipment

Explosion-proof equipment is contained in an enclosure that is capable of withstanding an internal explosion resulting from the ignition of a specified gas mixture. The enclosure prevents ignition of the specified gas mixture surrounding the enclosure. The enclosure must be flame-tight, meaning that the joints on the enclosure must be machined to a close tolerance to quench the flame. Furthermore, the path for venting must be long enough to cool hot gases to a safe temperature as they pass from the inside of the enclosure to the outside. Openings on the enclosure can be threaded joint type or ground joint type.

## Intrinsically Safe Equipment

Intrinsically safe equipment primarily is limited to instrumentation and some communications equipment. Intrinsically safe equipment is defined as that which is incapable of releasing sufficient electrical or thermal energy under normal or abnormal conditions to cause ignition of a specific flammable atmospheric mixture in its most easily ignitable concentration. The equipment and its associated wiring must be installed in such a way that positive separation between the intrinsically safe loop and other circuits is maintained at all times. This often is facilitated through the use of safety barriers at the hazardous/non-hazardous area interface.

## Purged and Pressurized Equipment

The use of purged or pressurized systems permits the use of electrical equipment that cannot be designed as explosion-proof. Switchgear and many large motors fall into this category. A purged or pressurized system requires a source of clean air or inert gas, a compressor to maintain the required pressure, and controls to prevent power from being applied before the unit is purged or pressurized.

## Switchgear and Controls for Hazardous Locations

A wide variety of explosion-proof electrical control equipment is available such as explosion-proof push-button stations, motor controls, and branch circuit breakers.

## Lighting Fixtures for Hazardous Locations

Functional illumination is the primary concern with hazardous location lighting. High-intensity, discharge type fixtures are being used for most new installations. Heat produced by the fixtures is an important factor in designing lighting for hazardous locations. Limiting temperatures have been established, and fixtures approved for Division 1 are marked clearly to indicate the maximum wattage of the lamp for which they are approved. Fixtures that can be used in Division 1 locations are explosion-proof and have a separate wiring compartment from the sealed lamp compartment. Separate sealing of these fixtures is not required.

To prevent breakage, lighting fixtures in Class I, Division 2 areas have to be protected by suitable guards or by mounting in strategic locations. Where lamps are of a size or type that may, under normal operating conditions, reach surface temperatures exceeding 80 percent of the ignition temperature in degrees Celsius of the gas or vapor involved, the fixtures must be of the type approved for Division 1. Otherwise, a type of lamp that is tested and found incapable of igniting the gas, or vapor, if the ignition temperature is not exceeded may be used. Therefore, fixtures used in Division 2 are generally of a vapor-tight design incorporating globe guards. Furthermore, these fixtures are gasketed to deny entry of flammable mixtures into the fixture.

Heat-producing equipment approved for installation in Class I locations of a particular group should not have any exposed surface that operates at a temperature in excess of the ignition temperature of the specific gas or vapor. Approved equipment must be marked to show the class, group, and operating temperature or temperature range referenced to 40°C ambient. The temperature range is identified by an identification number containing a "T" prefix. Note, the word "surface" pertains to equipment inside the explosion-proof enclosure and not to the enclosure itself.

Equipment that has been approved for a Division 1 location is permitted in a Division 2 location of the same class and group. Where specifically permitted by NEC Articles 501 through 503, general purpose enclosures may be installed in Division 2 locations if the equipment is non-incendive (does not constitute a source of ignition) under normal operating conditions.

## Motors and Generators for Hazardous Locations

Electric motors are needed to drive pumps, compressors, fans, blowers, etc., and their presence in hazardous locations at compressor stations is unavoidable. The selection of the proper motor becomes very important. The types of motors vary from open drip-proof to totally enclosed, fan-cooled motors. In Class I, Division 1 locations, only motors that are explosion-proof or of a totally enclosed, specifically pressurized type may be used.

Motors for use in Class I, Division 2 locations with sliding or switching contacts must be explosion-proof also. Open-type induction motors without arcing devices may be used in Class I, Division 2 locations if they meet the temperature restrictions mentioned earlier.

## Plugs and Receptacles for Portable Devices in Hazardous Locations

Plugs and receptacles for use in hazardous locations are made safe by the use of interlocks or delayed action, or a combination of the two. Receptacle contacts are interlocked with a switch, or the plug and receptacle are constructed so that any electrical arcs are confined to the inside of the explosion-proof chamber. Both designs are in wide use at compressor stations.

The use of portable equipment in conjunction with plugs and receptacles should be restricted as much as possible. When it is used, the NEC requires grounding by means of a separate grounding conductor in the portable cord.

## Wiring in Hazardous Locations

In Class I, Group D locations all conduit must be either rigid or intermediate metal conduit with at least five full tapered threads engaged. Explosion-proof, flexible connectors are available where lengths of flexible conduit are required. Type MI cable with termination fittings approved for location is acceptable also.

The NEC requires that sealing fittings filled with approved compounds be installed in conduits. Seals are necessary to limit volume and prevent an explosion from traveling through the conduit system. Seals also are used to prevent gases from traveling through the conduit systems from hazardous to non-hazardous locations.

Seals must be used in all of the following cases:
- Where the conduit enters an enclosure containing arcing devices or high temperature devices
- Where the conduit enters enclosures that house terminals, splices, or taps if the conduit is NPS 2 or larger
- Where the conduit leaves a Division 1 location or passes from a Division 2 location to a non-hazardous location.

In Class I, Division 1 locations multi-conductor cable in conduit may be considered as a single conductor if the cable cannot transmit gas through the cable core. If the cable can transmit gas through the core, the outer jacket must be removed so that sealing compound will surround individual insulated conductors.

Cables with continuous sheaths in Class I, Division 2 locations do not have to be sealed if they do not leak more than a sealing fitting with an approved sealing compound. However, if cables pass through the Division 2 location into a Division 1 location, they have to be sealed.

## Conduit and Conduit Systems

The NEC specifies the size of conduit to use depending upon the number, type, and size of the conductors to be installed therein. The proper size of conduit should be determined by tables provided in Chapter 9 of the Code.

Most compressor station wiring is located underground in some type of duct banks consisting of one or more conduits spaced close together with or without a concrete casing. (See Figure 310–1 of the NEC for recommendations.)

Manholes are necessary in a duct bank to permit the installation, removal, splicing, and rearrangement of the cables. Conduits enter the side of the manhole, and the cables are arranged within the manhole such that each is completely supported. The location of manholes is determined by the layout of the area to be supplied with power. Wherever a branch or lateral extends from the duct bank, there should be a manhole. Manholes should be spaced no more than 300 to 400 feet (90 to 120 metres) apart.

When manholes are not used, conduit junction boxes or pull boxes are used at points where several conduits intersect.

## Conductor Sizing

Four factors should be considered in determining the size of conductors in a compressor station: insulation temperature, mechanical strength, conduit size, and voltage drop.

The present maximum allowable, safe, current-carrying capacities of conductors are specified by the NEC in Tables 310–16 through 310–19 (see Appendix A). These ampacities are revised effective January 1, 1987, to account for variations of the ambient air and ground temperatures (Tables 310–20 to 310–30).

## Conductor Materials

Today only aluminum and copper conductors are in general use. Copper is by far the most universally accepted conductor for small wire sizes. Its mechanical and electrical properties make it ideal as a conducting medium. Not only does copper possess high conductivity, but it is ductile, fatigue resistant, and resistant to corrosion.

Aluminum often is used for large power distribution purposes. Aluminum is less expensive, but larger wire and raceway are required because of the reduced ampacity of aluminum.

Terminations for aluminum have caused considerable difficulty in the past because of creep, corrosion, presence of oxide film, and differences of coefficients of expansion of aluminum and other metal. Compression fittings with properly plated lugs must be used for aluminum conductors.

## Conductor Insulation

Currently the NEC lists as many as 15 insulation types for general use. The insulations most commonly used for compressor station applications are members of the thermoplastic family. Thermoplastic insulations can resist moisture, oils, and chemicals without protection. They are durable, tough, and lubricated with silicone, and are easy to handle and install.

Type TW and type THW wire are flame retardant and moisture resistant. In addition, type THW wire is also heat resistant. Type THHN and type THWN wires employ nylon jacket as outer covering and gradually are replacing other types of building wire. Their chief advantage is that their reduction in insulation thickness allows a greater number of conductors to be pulled into a given raceway. Type THHN wire is flame retardant and heat resistant, but its use is restricted to dry locations. Type THWN is moisture-resistant, and can be used in wet and dry locations. In larger sizes, type XHHW wire is used frequently. Its insulation is flame-retardant, cross-linked, synthetic polymer. It also can be used for wet and dry locations, but its maximum operating temperature has to be derated in wet locations.

## Conductor Voltage Rating

Low-voltage wire and cable currently being used for compressor station applications usually will carry 600-volt nominal ratings. Most control cable is rated at 600 volts; however, signal wire and instrumentation cable commonly are rated at 300 volts.

The NEC requires that all conductors in the same raceway, enclosure or cable have an insulation voltage rating equal to at least the maximum nominal circuit voltage rating of any conductor in the given raceway. Therefore, it is generally good practice to use wire and cable of the same voltage rating for the entire electrical system if practical; otherwise, conductors of high-voltage and low-voltage systems must not occupy the same wiring enclosure or pull box.

## Conductor Shielding

The object of shielding signal circuits is to protect the circuit from spurious electrical noise. This noise can be avoided by the use of shielded, twisted-pair construction. The shield grounding normally is done at one point only to eliminate the possibility of circulating currents. The use of steel raceways also will exclude induced magnetic fields generated by adjacent circuits.

## LIGHTING

Lighting system design may become quite complex. Manufacturers' literature and lighting handbooks should be consulted for specifics on illumination calculations and specific applications. A good reference is the *IES Lighting Handbook* copyrighted by the Illuminating Engineering Society of North America, 295 E. 47th St., New York, NY 10017; composed and printed by Waverly Press, Inc., Baltimore, MD. 1981.

## Indoor Lighting

In order to design a lighting system for any building, there are many engineering factors to consider such as number of fixtures, fixture types, fixture layout, and required fixture accessories.

Generally the selection of fixtures is specified by a catalog designation that is sufficient to obtain a fixture with the desired features, and would include all electrical characteristics such as voltage, wattage, and power factor. Selection of lighting equipment for hazardous locations should be based on NEC classifications commensurate with Class I and the appropriate Division for its environment.

General lighting should be supplemented with additional illumination in areas of a building where visual tasks must be accomplished. Where practical, indoor lighting should be downward with direct-type fixtures.

Emergency lighting always should be considered with indoor lighting system design. Fixtures used for emergency lighting may be individual, battery-operated units; incorporated into exit lights or directional signs; or they may be fixtures used with the general lighting but served from emergency circuits.

### Indoor Lighting Calculations

There are two approaches to providing the required illumination on tasks performed. If illumination is required at specific task locations, a Point Method may be used. However, it is usually more practical to design a lighting system to provide an average illumination level with a reasonable degree of uniformity throughout the area. The two methods most frequently used in the design of uniform indoor lighting are the Lumen Method and the Zonal-Cavity Method. Some lighting fixture manufacturers provide data applicable only to one of these methods, and the difference, therefore, has to be appreciated.

The Lumen Method is used in calculating the illumination that represents the average of all points on the work plane in an interior so that illumination is equal to luminous flux per unit area. Because not all lamp lumens will reach the work plane due to losses in luminaire and the room surfaces, an adjustment is made by introducing a coefficient of utilization (CU) and a light loss factor (LLF).

The Zonal-Cavity Method accounts for the effects of room proportions, luminaire suspension length, and the workplace height as well as the reflections of light from various surfaces. While the Zonal-Cavity Method of calculation is a bit more complex, the results are more accurate—frequently permitting the use of fewer luminaires.

### Outdoor Lighting

Many types of outdoor lighting equipment are available for use in compressor station yards including floodlights, street lights, protective lighting, and equipment lighting. Fixture selection is based on the construction and light output characteristics of the specific unit, and the physical characteristics of the area to be illuminated. Selection of outdoor lighting for hazardous locations should be based on the NEC requirements for the location. The use of sealed fixtures will result in the lowest maintenance cost and the highest average, maintained light level. Explosion-proof light fixtures are expensive and provide less illumination than a comparable sized standard fixture.

## Outdoor Lighting Calculations

The two most commonly used systems for floodlight calculations are the Point Method and the Beam-Lumen Method.

The Point Method involves the inverse-square law—the illumination on a given plane at a definite point, in footcandles or lux, is proportional to the candlepower (in candelas) of the source in the given direction, and inversely proportional to the square of the distance in feet (metres). Of course, when the surface on which illumination is to be determined is not normal to the light rays, the result will have to be multiplied by the cosine of the angle between the light ray and the perpendicular to the plane at that point. This method is valuable since it permits a visualization of the degree of lighting uniformity realized.

The Beam-Lumen Method is quite similar to the Lumen Method previously mentioned except that it must take into consideration the fact that luminaires usually are not directly above the surface, but instead are aimed at various angles to the surface.

## Lighting Illumination Levels

Recommended illumination levels for indoor and outdoor lighting applications at a compressor station are given in Table 2.

## MOTORS AND GENERATORS

Motors and generators should be selected with respect to the service conditions considering the application and environment. The majority of electric motors used at compressor stations are furnished by the supplier of the driven equipment, or they are a component of some type of packaged equipment.

## Induction Motors

Induction motors are by far the most common type motor used for drive applications at compressor stations. They are suitable for use on fans, compressors, pumps, and machine tool applications as well as almost any other drive application. They may be provided in a variety

## TABLE 2
### Lighting Illumination Levels
### at Compressor Stations

| Indoor Lighting | Footcandles on Task |
| --- | --- |
| Compressor Building | 20 |
| Auxiliary Building | 20 |
| Control Rooms | 30–50 |
| General Offices | 70–100 |
| Locker Rooms | 20 |
| Machine Shops | 50–100 |
| Meter and Regulator Building | 20–30 |
| Warehouse Buildings | 20–30 |
| Equipment Rooms | 20 |
| Pump House | 20 |

| Outdoor Lighting | |
| --- | --- |
| Building Exteriors | 5 |
| Building Entrances | 10 |
| Gate House | 10 |
| Catwalks | 2 |
| Roadways | 1 |
| Substations | 2 |
| Loading Platforms | 20 |
| Valve Yards | 10–20 |
| Storage Tanks | 1 |
| Outdoor Equipment | 10–20 |

of formats to meet torque requirements, speed variations, and special requirements where reversible drives are required.

Induction motors will operate successfully at the rated load with variations up to plus or minus 10 percent of the rated voltage, or up to plus or minus 5 percent of the rated frequency, and when the sum of the voltage and the frequency valuations do not exceed 10 percent above or below normal conditions. There will be a slight speed change when a load change occurs on an induction machine.

The current drawn by an induction motor consists of an in-phase component, which produces power, and an out-of-phase component needed to produce the magnetic field. The power factor of such a motor, therefore, always will be less than unity and gets worse under light loads, as does efficiency.

## Synchronous Motors

Unlike the induction motor the speed of a given synchronous motor is a function of the supply frequency. Thus, at 60 hertz a 2-pole synchronous machine will operate at 3 600 r/min exactly, a 4-pole machine at 1 800 r/min, a 6-pole at 1 200 r/min, etc.

Synchronous motors were first used because of their ability to raise the power factor of systems having large induction motor loads since the power factor of such systems frequently is quite low. When it is required that such a motor furnish part of the reactive current required by the system, the excitation is boosted beyond that required for the motor to operate at the unity power factor. (The excitation normally is provided by a direct current source.) In many cases, however, the synchronous motors cost less, and have higher efficiencies than the corresponding induction motors, especially in low-speed applications.

## Direct Current Motors

Direct current motors are the most common motor used for drive applications where the motor must operate without any normal alternating current source available, or where the application makes a direct current motor advantageous.

Direct current motors will operate successfully using the power supply selected for the basis of rating up to 110 percent of the rated armature voltage, provided the highest rated speed is not exceeded.

## Synchronous Generators

Synchronous generators, or alternators, operate in a very similar fashion to synchronous motors, but are used for power generator applications where constant voltage and constant frequency requirements must be met. Most on-site power generation equipment found at compressor stations is provided by some form of engine-driven, synchronous generator.

A synchronous generator will operate successfully at the rated kilovolt-amperes, frequency, and power factor with a variation in the output voltage of up to plus or minus 5 percent of the rated voltage.

## Induction Generators

An induction motor operates as a generator when spun above its synchronous speed with respect to the frequency of the power line to which it is connected. Because it takes reactive power for excitation, the induction generator generally is incapable of stand-alone operation, and cannot be used as an alternate power source during utility outages. Also, since the reactive power take is increased from the electric utility, the station power factor will be lower than that without on-site generation or with an alternator-based design.

## Motor and Generator Enclosures

Generally the frame of a motor or generator serves three distinct purposes:
- Transmits the torque to the motor supports
- Serves as a means of guiding the cooling medium
- Shields electrified (live) and moving parts from human contact and from injury caused by weather exposure or by falling objects.

For general-purpose motors, cast iron frames normally are used with cooling air admitted at each end and discharged from the midpoint of the frame after passing over the back of the core. Large machines most often have welded steel frames.

Formerly distinct types of frames were used for open, drip-proof, splash-proof, and totally enclosed motors, but the recent trend has been to concentrate on the "protected" and totally enclosed types only. A modern protected machine frame provides protection against falling objects and dripping liquids, and still provides adequate openings for a free flow of ventilating air.

In recent years totally enclosed machines have a much wider field of use whenever dust, corrosion, wetness, or explosion risks are encountered. Higher production further lowered the costs and increased their popularity.

## Motor Controls

Motor controls for induction motors may be of the full-voltage or reduced-voltage type. Any polyphase induction motor may be started safely by applying full voltage at starting without harm to the motor. However, heavy starting currents drawn from the line by this method of starting may produce unacceptable voltage fluctuations in the electrical system.

When the starting current must be reduced, some method must be used to reduce the voltage impressed on the motor during starting. This is accomplished by using primary impedances, resistors, or autotransformers installed in series with the motor winding. It should be noted that the motor torque varies as the square of the impressed voltage. Therefore, it is the required starting torque that will establish the lower limit for the impressed voltage.

In instances where reduced-voltage starting is not necessary, across-the-line starting may be used. Across-the-line starting may be used with or without thermal overload protection. Controls for this type of starting also may be purchased that include a motor circuit switch. These units are known as combination starters.

Controls for multi-speed induction motors are similar to those furnished for single-speed induction motors. The major difference is that they must provide a means for making the necessary changes in the connections of the winding for different speeds. Where separate switches are used for the different connections, provisions should be made for mechanically or electrically interlocking the switches.

Control equipment for synchronous motors is the same as that for induction motors with the addition of the field control equipment. A synchronous motor is started as an induction motor and when it reaches near-synchronous speed, a field contactor is closed, applying excitation to the motor field.

Controls for starting direct current motors are available in both manual and automatic, across-the-line systems, and may be used with or without thermal overload protection.


## GROUNDING SYSTEM

Grounding helps to prevent accidents to personnel and damage by fire to property in case of lightning, breakdown between primary and secondary windings of transformers, or accidental contact between high- and low-voltage wires. A low-impedance ground path return is absolutely essential if line-to-ground faults in grounded systems are to be cleared by overcurrent devices.

The NEC requires that the ground resistance must not exceed 3 ohms for water pipe grounds and 25 ohms for driven or buried grounds.

At compressor stations the ground reference usually is achieved with driven ground rods as the grounding electrode. Reduction of ground resistance in cases where it exceeds the allowable value of 25 ohms can be accomplished by the use of larger diameter rods, longer rods, connecting rods in parallel, or chemical treatment of the soil.

Ground rods used for grounding electrodes must be bonded together permanently. This is achieved with cable buried in the earth to form a loop. Station equipment is bonded to the ground loop and grounding electrodes. All equipment grounding connections should be made to this loop.

Electrical equipment and enclosures are grounded by carrying a separate grounding conductor, or alternatively by using the conductor support, raceway, or sheath of a metal-sheathed cable. The equipment grounding conductor always should be carried back to the initiating point of the feeders. This is usually the switchgear or motor control center. The equipment grounding conductor should be connected to the ground bus of this gear or its equal. Where a grounded conductor (the neutral) is bonded properly to the metallic service equipment, any line-to-ground fault from an ungrounded conductor will be able to follow a low impedance path to the grounded conductor.

Conductors used for grounding are ordinarily copper or aluminum cable. Aluminum or insulated copper conductors should be used where the soil is corrosive. Grounding conductors exposed above ground and subject to corrosion should be insulated, or of some material not subject to corrosion.

Grounding bushings, bonding jumpers, and approved grounding fittings are required when raceways are used as the ground return path.

Special attention should be given to the methods of grounding in hazardous locations to ensure that the requirements of the NEC, Articles 501 and 250, are met. The Institute of Electrical and Electronic Engineers (IEEE) Standard No. 80 provides excellent guidelines for equipment grounding in both hazardous and non-hazardous locations. (See Appendix A.)

## ELECTRICAL CODES AND STANDARDS

There are many codes and standards that apply to electrical systems and equipment with regard to the design, construction, operation, and maintenance. The Appendix, Codes and Standards, should be referenced for a tabulation of codes and standards applicable to compressor stations.

# INSTRUMENTATION AND CONTROLS

Control and instrumentation systems provide a means of regulating and monitoring compressor station day-to-day operations. Controls include any electrical or mechanical devices used to regulate or control station equipment. They are used to start and stop engines, regulate gas and air pressure, control liquid levels in lube oil tanks, control valve positions, and perform many other functions. The effectiveness of controls is monitored with instrumentation, which includes any device utilized to meter and monitor station equipment. Examples include flow meters, pressure gauges, and liquid-level indicators which can be found throughout the station. Because controls and instrumentation play a vital role in operations, proper selection and design of these systems are critical in maintaining a safe and efficient operation. The purpose of this discussion is to provide an overview of the various types of controls and instrumentation available, and to provide guidelines in the design and selection of this equipment.

## CONTROL PHILOSOPHY

The wide variation of compressor station size, complexity, type, and yearly utilization affects the sophistication and configuration of control systems selected.

Generally compressor stations can be subdivided into four types, or modes, of operation. The classification of each mode depends upon the degree of automation. Usually as the degree of automation increases, the sophistication of the controls and instrumentation increases also. The following is a brief description of each mode and the level of sophistication of its controls.

**NOTE:** *Instrumentation and Controls also is discussed in the Gas Control/Automation and Telecommunications Book, Transmission Volume, GEOP.*

### Attended Manual

The attended manual mode is the simplest of all modes. Some or all of the operating parameters may be set individually and controlled by the operating personnel. This mode of operation usually requires continuous supervision by station personnel to maintain the equipment within safe, efficient operating limits. Safety devices usually remain active in this mode to initiate automatic shutdown of equipment if a critical limit should be exceeded. Sequencing of valve openings and closings, and equipment starts and stops may be done manually or semiautomatically.

### Attended Automatic

The attended automatic mode relies on the control equipment to regulate various operating parameters, and to maintain the equipment within safe operating limits. Some or all of the operating parameter "set points" may be set by the operating personnel. Safety shutdown devices are active in this mode in the event that the control equipment fails. Valve and equipment sequencing usually are handled by automatic control equipment.

## Unattended Automatic

In the unattended automatic mode, control equipment completely regulates the station engine operation. Most operating parameters are set and maintained from a central location, frequently by a computer. Data logging is performed automatically by recording equipment to maintain records of system characteristics and equipment usage, as well as equipment malfunctions.

## Remote Automatic

The remote automatic mode is the most complex means of control. In addition to the complete control as described in the unattended automatic mode, there exists the ability to set and/or read operating parameters and control equipment from a remote location. This is done by using a supervisory communications system.

## CONTROL REQUIREMENTS

As discussed above, the complexity of the controls varies with the amount of automatic operation required and the mode of operation. They may range from simple, manually operated actuators to remotely controlled, computerized systems with trend logging and diagnostic capabilities. Regardless of complexity, control systems should be modular in construction to facilitate expansion and ease of maintenance. Controls may be mechanically, pneumatically, or electronically operated or activated. Mechanical and pneumatic controls are well suited for simple, on/off control tasks and sequencing. Controls of this nature are relatively inexpensive, and can be interfaced easily with engine/compressor equipment. They are also advantageous in that they can be used safely in locations classified as having hazardous atmospheres.

Electrical and electronic control equipment are utilized more often in complex applications where regulation of equipment requires more than simple on/off control, or where it is not feasible to transmit signals via long air lines to controlled equipment. Electric and electronic controls are advantageous because they can be interfaced directly to a programmable controller or a microcomputer, providing flexibility in operation and ease of expansion.

Depending upon the application, electric and electronic controls can be more costly than mechanical and pneumatic controls. Shorts or sparks generated accidently by equipment may start fires or cause explosions in classified locations; therefore, these controls must be installed using intrinsically safe equipment – or in purged or explosion-proof enclosures, which significantly increase costs.

Often the most efficient and cost-effective control system is a combined system, which includes components of all types of control equipment. Electronic and electric controls are interfaced with pneumatically and mechanically operated controls to produce safe, extremely accurate control systems at a minimum possible cost. Table 3 gives a listing of the most common monitor and control parameters used in a compressor station. Also shown in the Table is the device that is used to sense an abnormal condition or provide a means of controlling certain operations.

## Prime Mover Controls

Prime mover controls include all equipment that regulates the operation of the prime mover. These controls include vibration, speed, ignition, loading, and safety devices installed within, on, or at remote locations from the engine. Those that are critical to the safe operation of equipment sometimes are monitored by redundant equipment with at least one set being failsafe. A failsafe operation protects equipment and personnel in the event of a malfunction. It usually results in immediate shutdown of the equipment. A complete discussion of engine protection devices is provided under the subject of Safety and Security.

In addition to monitoring and adjusting prime mover operating conditions, the control equipment usually ensures correct sequencing of certain steps such as the start/stop sequence. For a prime mover, start/stop sequencing involves the sequential operation of pre-lube, starting power, ignition, fuel, and loading devices. Interlocks should be provided to keep manual operations from occurring in an improper sequence.

For illustrative purposes a typical start/stop sequence for an automated, natural gas-powered, turbocharged, 10-cylinder engine is described below. In this example the engine is a reciprocating unit with a speed of 300 r/min. Generally the start sequence is initiated by a start command from a control panel in the compressor or auxiliary building, or at another remote location. The sequence automatically proceeds as follows:

1. The yard valves are checked and verified to be in the proper position – vent valve open; bypass valve open; and suction and discharge valves closed.
2. The engine and turbocharger pre-lube pumps are started, providing lubrication to the engine and turbocharger bearings before an actual start occurs.
3. After the oil pressure reaches 15 lbf/in² (100 kPa), the starting

## TABLE 3
### Compressor Station Monitor and Control Parameters

| Operation | Condition | Classification | Device |
|---|---|---|---|
| Auxiliary Generator | Main Power Failure | A-C | Relay Operated Starting Equipment |
| Battery | Low Voltage | A | Low Voltage (DC) Relay |
| Bearing Temperature | High | A | Temperature Switch |
| Compressor Loading Valves | Open or Close | C | Valve Operator |
| Compressor Vibration | Excessive | A | Vibration Switch |
| Control of Compressor Unit | Remote or Local | S | Manual Control Switch |
| Coolant Level | Low | A | Float Switch |
| Coolant Radiator Vibration | Excessive | A | Vibration Switch |
| Coolant Temperature | High | A | Temperature Switch |
| Current | High | A | Circuit Breaker |
| Discharge Pressure | High or Low | A | Pressure Limit Switch |
| Discharge Pressure | Raise or Lower | C | Controller/Throttling Valve |
| Discharge Temperature | High | A | Temperature Switch |
| Emergency | Stop | A-C | Emergency Shutdown System |
| Engine | Start | C | Starting Air Valve/Pilot/ Idle Speed Controller |
| Engine | Stop | S | Fuel Line Valve/ Ignition Switch |
| Engine Speed | High or Low | C | Governor |
| Engine Speed | Increase or Decrease | C | Controller |
| Fire | Glowing | A | Fire Detector |
| Flow Rate | Absolute Value | S | Flow Meter/Transmitter |
| Hazardous Gas | Explosive Condition | A | Gas Detector |
| Instrument Air | Low or Failure | A | Pressure Switch |
| Lube Oil Level | Low | A | Float Switch |
| Lube Oil Pressure | Low | A | Pressure Switch |
| Lube Oil Temperature | High | A | Temperature Switch |
| Motor | Start | C | Starting Control |
| Motor | Stop | S | On-Off Switch |
| Motor Winding Temperature | High | A | Temperature Switch |
| Scrubber Liquid Level | High | A | Float Switch |
| Starting Air | Low or Failure | A | Pressure Switch |
| Start Sequence Failure | Abnormal | A | Start Cycle Timer Switch |
| Turbine Speed | High | A | Overspeed Switch |
| Turbine Speed | Increase or Decrease | C | Controller |
| Turbine Vibration | Excessive | A | Vibration Monitor |
| Voltage | Low | A | Circuit Breaker |

Key to Classifications:

A–Denotes function in which, when an abnormal operating condition exists, instrument takes action, sounds alarm, and indicates (usually by light signal on panel) the presence of the abnormal condition.

C–Denotes operation can be controlled automatically by instruments that adjust to changes in operating variables as necessary.

S–Denotes function in which intelligence is both measured and/or recorded and is also transmitted to remote point for reading; supervision is either remote or local or both, by either automatic or manual adjustment.

*(Reference: American Gas Association 1974 Operating Section Proceedings, Catalog No. X50674; "Surveillance for Unattended Natural Gas Compressor Stations," Page T-36)*

air and starting jet-assist valves are opened, and the turbocharger begins to rotate.

4. After an 8- to 10-second delay, starting air is delivered sequentially to the power cylinders through the starting air distributor. Starting air enters the cylinders and forces the pistons downward, which causes the engine crankshaft to rotate.

5. When the engine speed reaches 40 r/min, the pneumatic ignition switch is energized. After a 4-second delay, the ignition system is activated electrically so that the spark plugs are ready to fire.

6. If all engine shutdown devices are inactive and a minimum scavenging air pressure of 1 inch Hg (3 kPa) is reached, the fuel regulator is opened to supply fuel gas to the engine at 18 lbf/in² (120 kPa).

7. When the engine speed reaches 125 r/min, the fuel pressure regulator gradually increases the engine fuel pressure to 70 lbf/in² (480 kPa). At this time the starting air valve is closed; the pre-lube pumps are stopped; the vent valve is closed; and the suction and discharge valves are opened. The low oil pressure, shutdown device is activated also.

8. The engine will operate at idle speed (200 r/min) for 2 minutes before the bypass valve is closed and loading occurs. At this time the engine governor is placed into service, and the engine gradually accelerates to 300 r/min.

9. Compressor clearance pockets are adjusted automatically to obtain the optimum engine torque.

A normal automatic stop sequence is much simpler than the start sequence and proceeds as follows:

1. Having received a stop command, the engine is unloaded by opening the bypass valve and closing the suction and discharge valves. The engine speed automatically changes to idle speed without adjusting the clearance pockets.

2. After the engine has run in the idle position for 3 minutes, the fuel gas is shut off to the engine and the intake manafold vented to the atmosphere. After the engine speed drops below 40 r/min, the ignition automatically shuts off.

3. The post-lube pump is started and operates for 3 minutes to provide lubrication to the bearings during the cooldown.


## Compressor Controls

Controls required to regulate compressors are just as important as prime mover controls. The most often regulated and monitored parameters are lubricant flow rate and pressure; cooling water temperature; bearing lube oil pressure, temperature, and level; bearing temperature; vibration level; cylinder liner temperature and valve temperature; gas temperature and pressure; and compressor surge in centrifugal compressors. When a critical set point for any one of these parameters exceeds a predetermined value for safe operation, safety devices will automatically shutdown the engine/compressor unit.

Control equipment can be utilized also to calculate and control the loading on the compressor by means of a clearance pocket adjustment and/or bypass valve arrangement. Usually the control equipment for both the engine and the compressor is located in the same cabinet since they must interact with each other during the operation.

In some applications it also is desirable to incorporate engine speed variations as a means of controlling the gas flow rate. For instance, a controller could be used to sense suction gas pressure and regulate the speed over a suitable range according to some predetermined suction pressure set point.

## Valve Controls

The control of valve operations is an integral part of the task of controlling engines and compressors. Dependability is critical as these valves must be relied upon to protect station equipment and to route the gas through the station. Failure of any of these valves or their control equipment could render a station useless. For example, relief valves in the yard piping protect the compressors, piping, and associated equipment from overpressure.

The valve position is controlled hydraulically, pneumatically, or electrically by valve actuators. Usually the position of a valve is determined through the use of limit switches or potentiometers. For on/off applications, limit switches indicate the valve position by making or breaking circuits between the valve and the controller. Potentiometers transmit an electrical signal to the controller which is proportional to the valve position.

When multiple valves are controlled by an automatic means, valve sequencing becomes important. Valve sequencing is the process by which valves are actuated sequentially by an automatic means to perform an operating function. The actual valve sequence can be determined by current valve positions or by timers built into the control system. Typical valve sequencing operations include the changing of yard valve positions in preparation for bringing units on line, or closing station block valves and opening vent valves during an emergency shutdown sequence. Regardless of the application, all automatic sequencing operations should provide interlocks to prevent accidental valve operation if a partial or complete failure of the control system should occur.

### Auxiliary Equipment

Controls for auxiliary equipment to the prime movers and compressors such as electrical generators, batteries and charging equipment, air compressors, cooling fans and pumps, and building heating and cooling equipment, are similar to prime mover and compressor controls. Auxiliary controls can be integrated with prime mover controls or have a separate control system of their own.

Electrical generators may be driven by the same type of prime mover used to drive the gas compressors; consequently, the controls for these units would be similar to those used on the gas compressor prime mover.

Battery power supplies and chargers utilize controls to maintain the proper voltage in the system.

Usually air compressors are equipped with controls to maintain a specified pressure in the air storage system.

Cooling fans and pumps often require control equipment to monitor and maintain such parameters as coolant temperature, coolant pressure, coolant flow, oil temperature, oil pressure, and oil flow.

Environmental heating and cooling equipment may be necessary to maintain a specified temperature, particularly in buildings where computer equipment is located.


## INSTRUMENTATION

Instrumentation is the interface between the equipment and the operator. It is used to display and/or record the parameters upon which the control equipment acts.


### Indicating Equipment

A control system would not be complete without some form of indicating equipment. A typical indication is provided by an alarm status panel with visual and audible alerts. The indication can be as simple as a gauge or sight glass, or as sophisticated as a cathode-ray tube (CRT) color display. Each situation dictates the type of instrumentation that is needed.

Environmental conditions usually determine the type of instrumentation used. Pneumatically or electrically operated devices have been used traditionally in areas where weather, vibration, or temperature extremes are severe. Recent developments in electronic equipment, however, have made it useful for these locations. Electronic equipment is advantageous because information can be stored readily and used in trending and diagnostics.

## Recording Equipment

Permanent records of certain operating information, such as alarm information or specific operating conditions, should be maintained. This information may be used for a variety of purposes such as maintenance of station equipment, preparation of reports, troubleshooting, or even billing.

There are numerous types of recording equipment available such as disks and roll chart recorders, magnetic disks, printers, and electronic memories. Information stored on magnetic devices or in memories requires additional equipment such as displays, chart recorders, or printers to present the information in readable form. The advantages of magnetic or electronic storage are density of data and speed of retrieval.

More sophisticated devices require a special operating environment. Magnetic equipment should not be used around high current electrical devices without proper shielding since magnetic fields generated around this equipment can destroy stored information and possibly damage the recording equipment. Mechanical devices, such as chart recorders, are better suited for these areas.

## Supervisory Control and Data Acquisition

Encoded information and commands are exchanged between the operator in the central control center and the remote station being controlled by a communication link. The link itself may be a telephone line, radio, or microwave, to mention a few; the equipment converting the data into a form suitable for transmission over the link and for display after it is received completes the Supervisory Control and Data Acquisition (SCADA) system.

Any amount of information can be transferred, ranging from a single parameter to a complete set of operating data for the station. Typical commands, alarms, status, and data transmitted between the station and the control location are listed below.

### Commands

- Start/stop of engine/compressor unit
- Change of operating set points
- Start/stop of station

### Alarms

- Fire
- Gas detection
- Intrusion by unauthorized personnel
- High gas pressure

* Low gas pressure
* High gas temperature
* Low gas temperature
* Power failure
* Prime mover failure
* Operator check-in
* Valve failure

**Status**

* Attended
* Unattended
* Station on line
* Unit on line
* Unit off line
* Unit available

**Data**

* Temperatures
* Pressures
* Set points
* Flows
* Electrical Power Usage
* Engine Power

A thorough discussion of the technology of remote control, data acquisition, and telemetering is contained in the Gas Control Automation and Telecommunications Part of the Gas Engineering and Operating Practices series. A modern control room for a large compressor station is shown in Figure 67.


# LIQUID STORAGE

Facilities for storing lubricating oils, waste oils, water, methanol, ethylene glycol, and other liquid hydrocarbons should be considered in the design of a compressor station, and should include tanks and other appurtenances.


## LIQUID STORAGE TANKS

All liquid storage tanks should be equipped with suitable fill and drain connections, sight gauges, shutoff and control valves, maintenance valves, safety valves, vacuum breakers, relief valves, and manways. Tanks operating at 15 lbf/in² (100 kPa) or greater pressure should be fabricated

*Figure 67.* Modern control panel for reciprocating engine/compressor
unit

in accordance with the American Society of Mechanical Engineers
(ASME) *Boiler and Pressure Vessel Code*, Section VIII (see Appendix
A). Fill piping should extend to the bottom of the tank, and a grounding system should be incorporated on all tanks and fill connnections
to control static electricity. Tank capacity requirements are determined
by the economics of bulk purchases and the operating need for a complete product exchange. Thermostatically controlled heat to maintain
minimum liquid temperature also may be required in certain climates
and environments.

Catwalks and stairs should meet Occupational Safety and Health
Act (OSHA), Section 1910.23 requirements (see Appendix A).

## Lube Oil Maintenance and Antifreeze Storage

Lube oil maintenance and antifreeze coolant storage tanks should
be installed to permit storage of the lube oil or coolant that is drained
from an engine or compressor during maintenance or overhaul. These

tanks may be designed to transfer the fluids by gravity, pressure, or a pump. They should have the capacity to drain the largest compressor cooling and lubricating system including piping, pumps, filter housings, and coolers. In addition, a freeboard margin should be provided when the tanks are sized.

## Jacket Water Supply and Maintenance System

A jacket water drain-and-fill system should interconnect the low point of each cooling water system with a self-priming antifreeze pump. Provisions should be made to drain the engine power cylinder and water jackets, as well as all piping, coolers, surge tanks, and compressor cylinder jackets into the maintenance tanks.

## Hydrocarbon Storage Tanks

Hydrocarbon storage tanks are required where inlet scrubbers or separators remove liquid hydrocarbons from the gas stream, or at dehydration plants where liquid hydrocarbons are removed together with the entrained water. Lead time for disposition of such collected liquid hydrocarbons will govern the volume of tankage required.

The hydrocarbon storage tanks should be designed in accordance with The American Petroleum Institute (API) Standard 650 (see Appendix A). The tanks should be equipped with flame arrestors as required by API Standard 2000 and the National Fire Protection Association (NFPA) Code 395 (see Appendix A).

## SPILL PREVENTION

Establishment of spill prevention control and countermeasure (SPCC) plans have to be provided in many oil or oil-related waste storage facilities. The exact requirements are outlined in the Code of Federal Regulations, Title 40 – Protection of the Environment, Part 112 – Oil Pollution Prevention. The SPCC plan has to include spill prevention procedures, monitoring, emergency action, and notification after the occurrence. These regulations are applicable to facilities that, because of their location, could reasonably be expected to discharge oil in harmful quantities into or upon the navigable waters or contiguous zone of the United States. This includes most surface water and wet lands.

Small capacity storage facilities – that is, those consisting of a single above-ground storage tank with a capacity of 660 gallons (2 498 litres) or less, or an above-ground facility with a total capacity of 1 320 gallons (4 997 litres) or less, or an underground facility with a capacity of 42 000 gallons (158 987 litres) or less – are exempt from the 40 CFR, Part 112 requirements.

## Secondary Containment

Secondary containment, when required for flammable products and oil product storage tanks, should include one of the following methods: diking, curbing, diversion or retention ponds, or sorbent materials. It is recommended that containment capacity be 110 percent of the capacity of the largest tank within the area to allow for precipitation, and that the containment be impervious. State and local codes should be consulted as some are more stringent than the federal code.

## Safety

All storage tanks should be identified clearly, and fire control or personal protective equipment should be provided in accordance with OSHA standards.

**NOTE:** *Instructional signs for loading and unloading storage tanks should be erected and grounding/bonding cables should be installed if applicable. If non-company personnel are to load or unload the storage tanks, consider locating the tanks to avoid necessity for plant entry. If not practical, attempt to minimize exposure of plant equipment to vehicular traffic.*

# WATER AND WASTE SYSTEMS

Water supply and waste disposal systems are essential elements in the design of a compressor station. Additional design requirements are imposed if there are no municipal services available. General design considerations and options are listed for these additional requirements.

## WATER SUPPLY

The local health authority should be consulted for regulations and available sources of water supply when selecting a site. The American National Standards Institute Plumbing Code, ANSI A40.8, and the Building Officials Conference of America (BOCA) Basic Building Code provide additional information regarding the design of water supply systems. Water samples should be submitted for analysis on a regular basis to the appropriate health authority. (See Appendix A.)

### Potable Water

The potable water quality should meet the chemical and bacteriological requirements of the United States Public Health Service Drinking Water Standards. Chlorination is the most widely accepted method of meeting these requirements. Chlorine and phenolic tastes can be controlled by activated carbon filtering.

### Utility Water

Utility water should not be corrosive, or contain substances which would result in damage to equipment. Utility water may require filtration and treatment through an environmentally acceptable system. Competent advice on water treatment may be secured from a consultant, but frequently is offered by the chemical and equipment suppliers.

### Fire Protection

The fire protection systems should meet NFPA standards (see Appendix A). Fire Underwriters and Fire Insurance Service Offices can be consulted for design recommendations. A good general reference is the *Fire Protection Handbook* which may be obtained through the National Fire Protection Association. (See Appendix A.)

Combustible material should not be stored in a compressor station except as necessary for daily operations. For the safety of personnel fire protection equipment must include gas, smoke, and fire detectors capable of sounding an alarm or activating an emergency shutdown (ESD) system. Chemical and foam systems can control fire effectively in most instances where gas can be evacuated or eliminated as the source of fuel.

A wet system may be preferred for protection of company housing or a warehouse. A sprinkler system for inside control or fire hose with a hydrant system for outside control are recommended. The minimum water supply requirement is 500 gal/min at 20 lbf/in$^2$ (31.5 L/s at 140 kPa) residual; however, twice the rate at the same residual is recommended. Pumps should have a positive suction pressure, and the water supply (well, stream, lake, or reservoir) should be protected from freezing at the intake.

A water stream extinguishes fires by cooling. It is the most effective means of removing heat from ordinary combustible materials used in building construction. Removal of heat eventually stops the release of combustible vapors and gases from these materials, which are involved directly in the burning process. Of course, fires in closed spaces may be put out by oxygen dilution, as hydrocarbon gases and vapors usually will not burn when the oxygen level is less than 15 percent. This is the mechanism involved in carbon dioxide flooding systems. (When a portable $CO_2$ system is used, the discharge plume entrains air, the residual velocity of which helps to blow the flame out while cooling the surfaces to prevent re-ignition.) Fuel removal is one of the most effective ways to fight a fire, and can be accomplished literally by shutting off the gas supply by a valve or by covering the liquid and solid burning fuels with a blanket of fire-fighting foam.

More recently, extinguishment by chemical flame inhibition has been gaining attention. The flames are extinguished because the extinguishing agent interferes with hydroxil formation needed for flame propagation. Bicarbonate of soda is used in many applications, but halogenated hydrocarbons (the Halon family) do not leave the residue after use and exhibit low toxicity. Inhibition of flames is very fast, and the method has been used to suppress explosions after the occurrence of ignition of a flammable mixture. However, the method is not very effective in combating glowing fires. The glowing fire problem is best handled by water sprinklers.

## SANITARY SYSTEM

Economics often dictate an on-site sewage disposal system such as septic, lagoon, mound, or mechanical treatment.

A sewage disposal system should be designed properly, considering the local soil conditions, to meet local health authority requirements. Regular maintenance is imperative for satisfactory operation. After determining the quantity of effluent to be treated, percolation tests should be conducted to determine the permeability of the soil and the area required for treatment. A study should be made regarding the potential for contamination of the water table. An evaluation of the site should be made for excessive slope, flooding, and proximity of wells, lakes, streams, or buildings. The local health authority also should be consulted for applicable regulations.

## PLANT WASTE

The Water Pollution Control Act prohibits the discharge of contaminated effluents into navigable waters without appropriate National Pollution Discharge Elimination System permits. Such discharge at a compressor station may include cooling water, floor drainage, sanitary sewerage, hydrostatic test water, treated jacket water, and other process wastes. There are strict regulations governing the discharge of pollutants such as chromium, lead, and mercury. Environmental Protection Agency (EPA) regulations should be reviewed carefully, with particular attention paid to the Clean Water Act of 1977 (Public Law 95–217), and the necessary permits should be obtained before construction is started (see Appendix A).

## Waste Disposal

The Resource Conservation and Recovery Act of 1976 (40 CFR, Parts 260 through 265) deals with discarded materials which are toxic, ignitable, reactive, or corrosive (see EPA, Appendix). All facilities generating or disposing hazardous wastes in excess of 1 000 kilograms (2 200 pounds) per month are required to file as a waste generator. On-site storage by a waste generator is limited to 90 days. Burial, evaporation, or burning is prohibited. Hazardous waste treatment should be done by a licensed facility. Therefore, frequent and manifested transfer by licensed waste haulers to an EPA-designated Hazardous Waste Management Facility are necessary as specified by the Act.

# Chapter 3

# OPERATIONS

This chapter describes the various types of compressor station operations, and the duties of personnel assigned to operate and maintain station facilities.

## PERSONNEL REQUIREMENTS

Compressor station personnel requirements vary with the type and size of the station. There are four basic types of station operations:

- Attended manual
- Attended automatic
- Unattended automatic
- Remote automatic

For larger stations (8 to 10 units) the operating personnel can be divided into eight classifications as noted below and illustrated in Figure 68.

- Station Superintendent is responsible for the administration, operation, and maintenance of the entire facility.
- Supervisor assists the superintendent as the front line supervisor, directing the activities of operating and maintenance personnel.
- Clerk performs clerical duties including cost accounting, material accounting, operating statistics, historical records, and general office correspondence.
- Shift Operators are responsible for around-the-clock operation of the station's mechanical equipment.
- Operators work under the direction of the shift operator on an around-the-clock schedule. Included in the duties of these operators are station housekeeping, cleaning, and minor maintenance.

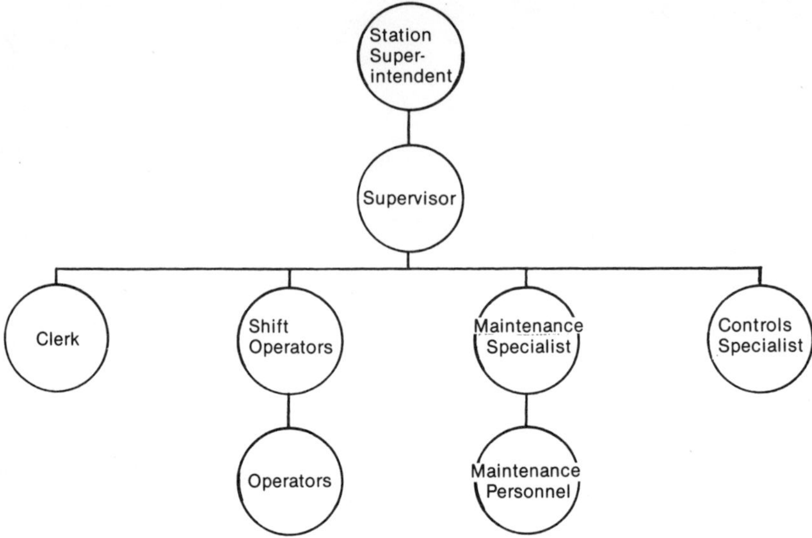


*Figure 68.* **Typical compressor station operational personnel**

- Maintenance Specialist is skilled in mechanical maintenance and repair, and serves as leader of the other employees in the maintenance crew.
- Controls Specialist is qualified technically to maintain the automatic control equipment.
- Maintenance Personnel are experienced, in or are being trained for, engine maintenance and repair, and maintaining the station equipment in a good mechanical condition. Maintenance personnel work under the direct leadership of the maintenance specialist.

The following examples describe the four types of compressor stations, and show the typical manpower requirements for a station with 8 to 10 units. Table 4 summarizes the number and classification of full-time employees for each example.

## ATTENDED MANUAL

At an attended manual station all functions, such as engine starting and stopping, loading and unloading, monitoring operating parameters, recording engine data and handling emergencies are done manually.

Since this type of station is designed with a minimum of automatic

## TABLE 4
## Suggested Employee Complement for Station with 8 to 10 Units

Personnel Requirement Summary

| Type of Station | Station Super-intendent | Supervisor | Clerk | Shift Operators | Operators | Maintenance Specialist | Maintenance Personnel | Controls Specialist | TOTAL |
|---|---|---|---|---|---|---|---|---|---|
| Attended Manual | 1 | 1 | 1 | 4 | 8 | 1 | 7 | – | 23 |
| Attended Automatic | 1 | 1 | 1 | 4 | – | 1 | 9 | 1 | 18 |
| Unattended Automatic | 1 | 1 | 1 | – | – | 1 | 9 | 1 | 14 |
| Remote Automatic | 1* | 1* | 1* | – | – | – | 7 | 1 | 11 |

*Performs duties for several stations at a central location.

control devices and work aids, it requires the largest personnel complement of the four cases under consideration – 23 regular employees plus short-term employees.

## ATTENDED AUTOMATIC

An attended automatic compressor station is one in which automatic devices and systems have been installed to perform some, but not all, of the operating functions.

A regular complement of 18 employees, plus temporary, short-term employees to provide replacements for job vacancies, should satisfy the personnel requirements of an attended automatic station of this size.

## UNATTENDED AUTOMATIC

An unattended automatic station is one which is attended only during the regular workday with no personnel on duty during the night, weekends, and holidays. Shift operating personnel are not required at this type of station. Engine controls, emergency systems, data recording, and all other station functions are monitored and controlled by automatic equipment.

A complement of 14 regular employees – plus temporary, short-term employees – should satisfy the personnel requirements of an unattended automatic station of this size.

## REMOTE AUTOMATIC

A remote automatic station is similar to an unattended automatic station except that it has additional equipment installed to telemeter operating data and control functions to a location remote from the immediate station site.

This type of station, often referred to as a satellite station, is operated from a control center some distance away. A complement of 11 full-time employees – plus temporary, short-term employees – is required to perform the duties related to the station operation, but only eight of these employees devote 100 percent of their time directly to this location. Since the supervisors and the clerk devote only part of their time, a cost savings on labor can be realized.

# DUTIES OF PERSONNEL

Compressor station operations encompass all supporting activities to safely, economically, and reliably transport gas to the market or to a gas storage facility. It is the duty of the station superintendent and personnel to ensure the safe, efficient operation of station equipment through good housekeeping practices, and proper maintenance of records and reports.

## STATION SUPERINTENDENT'S RESPONSIBILITIES IN STATION OPERATIONS

The station superintendent is responsible for overseeing the entire station operation. His duties include the following:

- Supervising the administration of office procedures covering cost accounting, record keeping, financial transactions, communications, and community relations
- Administering personnel policies and maintaining good working relationships with the employees under his supervision as well as with other employees with whom he associates
- Supervising the training of operating and maintenance personnel
- Maintaining station security and safety practices to prevent damage to company property and personal injuries to employees
- Ensuring compliance with federal, state, and local rules and regulations governing station operations and related activities.

The station superintendent is the key person at the facility. He must have the necessary technical and administrative skills to direct his personnel in the performance of their duties. He is often the difference between an efficient, trouble-free station and a poorly run station.

## HOUSEKEEPING

Good housekeeping contributes to the successful operation of any facility. Accidents are reduced by eliminating conditions such as slippery walkway surfaces, foreign objects in walkways, cluttered work areas, and poorly maintained tools and work equipment. Equipment can be protected from contamination or fouling by dirt or dust through regular cleaning of building interiors. Good public relations, personal pride, and high morale among employees are fringe benefits resulting from good housekeeping practices.

## RECORDS AND REPORTS

Records and reports for compressor stations fall into two categories: equipment operating and maintenance records, and regulatory agency reports and records.

### Equipment Operating and Maintenance Records

Equipment operating records include charts, graphs, log sheets, and other recorded data that are used to determine the most efficient way to operate the equipment. Maintenance records include data developed during the routine checking and inspection of equipment, preventive maintenance work, routine engine overhauls, and emergency repairs. These records are used to predict equipment dependability, and to develop planned maintenance programs that will ensure a reliable operation.

Also, records are retained on file at a station to aid in the forecasting of cash flows, budgeting of station operating and maintenance costs, planning of work schedules, predicting of manpower requirements, and setting of long-range, operating goals.

Reports are submitted to company management to keep them informed of operating activities. Copies of reports pertaining to operation and maintenance, history, safety, and environmental records become part of the permanent station records.

### Regulatory Agency Reports

Complete records and reports must be prepared and filed as mandated by the rules and regulations issued by federal, state, and local agencies. These records often require detailed inspections and reports of the operating conditions within the station, as well as records testifying to environmental considerations, safety, training, and personal protection afforded the employees.

# PERSONNEL TRAINING

Personnel training has become an essential element in the operation of modern compressor station equipment. An effective training program includes aspects relating to the operation and repair of the equipment, as well as procedures for handling emergency situations.

## TRAINING PROCEDURES

Early in the history of the industry, little effort was made in training employees in the operation of compressor engines and associated equipment. The employee gained knowledge through long periods of operating experience and his own efforts toward self-education. The advent of sophisticated engine controls and unattended automatic stations brought with it, however, the need for formal technical training. Training programs can be handled in several ways, and can be as extensive as necessary to ensure the safe, efficient operation of the station. Such programs include the following:

- On-the-job training for new employees
- Self-study courses financed by the company
- Organized programs conducted by a training group within the company
- College credit courses subsidized by the company
- Programs conducted by organizations specialized in the technical training field
- Seminars on technical subjects conducted by factory representatives or consultants in specialized fields
- Technical classes conducted by colleges or universities, jointly sponsored by gas industry associations
- Roundtable conferences or workshops sponsored by gas associations.

A well-conducted training program provides qualified employees for the operation and maintenance of sophisticated equipment in modern stations, and establishes a manpower pool from which employees can be selected to fill higher level job openings in the future.

## STANDARD AND EMERGENCY OPERATING PROCEDURES

The Code of Federal Regulations Title 49, Part 192, (see Appendix A) established certain minimum federal operating standards for gas pipelines. Each pipeline company is required to establish a written operating and maintenance plan that meets the minimum federal standards. The plan is used as a guideline for training and directing employees in the administration and execution of correct operating procedures.

Included in the standards are emergency operating procedures for the compressor station which cover the following:

- Emergency action to be taken during emergencies such as fires, pipeline ruptures, and explosions
- Emergency action guidelines for use in training employees to handle emergencies
- Provisions for materials, equipment, and trained personnel to be utilized in an emergency
- Procedures for reporting gas leaks, pipeline failures and other interruptions of pipeline service
- Notifying company officials and regulating agencies of major emergencies
- Investigation of major system emergencies.

*Chapter 4*

# MAINTENANCE

Factors to be considered in developing and maintaining a sound compressor maintenance program include:
- Scheduling, which varies with the type and age of equipment
- Load factor, use factor, and operating environment
- Spare parts availability
- Personnel and training requirements
- Monitoring and inspection techniques.

## SCHEDULING

Maintenance scheduling is determined by company policy, and can be performed on the basis of fixed intervals or predictive failures.

### FIXED-INTERVAL SCHEDULING

A fixed-interval program schedules maintenance of equipment after a fixed period of time elapses or after the equipment has operated a predetermined number of hours. The interval length is determined by considering the operating history and reliability factors of the equipment. The duration of the interval can be extended significantly through continuous monitoring of critical equipment components, periodic equipment testing, and on-line repair of minor malfunctions.

Maintenance of gas turbines usually is performed on a fixed-interval basis according to the turbine manufacturer's recommendations.

### PREDICTIVE FAILURES SCHEDULING

Predictive maintenance scheduling is based on predicted component failures either as the result of observed digression from normal operating parameters or an evaluation of historical data. Major maintenance should be scheduled when test equipment and/or operating data indicate a major problem, or when certain parts will have approached their maximum life expectancy.

Reciprocating engine and compressor maintenance usually is scheduled on a predictive basis.

## EMERGENCY MAINTENANCE AND SCHEDULING

Emergency maintenance is performed after a failure has occurred. Usually only the component which has failed will be repaired or replaced.

## PERSONNEL REQUIREMENTS

Maintenance and repair of compressor station equipment are performed by both company and contracted personnel.

### Company Personnel

In most cases company personnel are utilized to perform routine maintenance and repairs. The number and classification of company personnel will be determined by the type of equipment and the extent of maintenance to be performed. Usually a crew will consist of a maintenance specialist and one to six maintenance personnel.

### Contracted Personnel

In the case of major maintenance or a major component failure where specialized skill is needed, the equipment manufacturer's personnel or personnel from a service organization may be contracted to make repairs on-site. If repairs or maintenance cannot be performed on-site, either the component or the entire piece of equipment is removed and shipped to a maintenance facility for inspection, maintenance, and repair.

Examples of jobs done by contracted labor include regrouting of compressor foundations, major repairs such as broken main frames or crankshafts, and repair of turbines.

# MAINTENANCE REQUIREMENTS

Monitoring engine operating data and routine investigation of engine performance are key factors in a maintenance program. Engine operating data should be recorded on an appropriate log sheet at least daily, and perhaps more frequently, by the station operator. The operator should know the operating parameters for all types of engines, both turbine and reciprocating, at the compressor station. When any operating data approaches set limits or when there is a drastic change in any of the operating data, the operator should notify the supervisor immediately. Figures 69, 70, and 71 illustrate typical log sheets for reciprocating and turbine engines.

When computers are used to record operating data, the same philosophy should apply. Parameters should be programmed into the computer so that alarms are activated when any operating data exceeds these parameters. The computer can scan all operating data rapidly; therefore, a more accurate trend of changes in operating data can be established.

## MAINTAINING PRIME MOVERS

Prime movers can be divided into two major divisions: reciprocating and gas turbine. In some cases electric motors are used as prime movers, but since there are so few in service in the industry, they will be omitted from further discussion.

### Reciprocating Engines

Reciprocating engines can be protected from catastrophic failures and maintained at top operating efficiencies by installing engine shutdown devices and adopting a preventative maintenance program.

Engine shutdown devices consist of a system of sensors that monitor critical engine parameters (for example, lube oil temperature, jacket water temperature, engine speed, vibration, etc.). When a sensor detects an abnormal condition, it activates the shutdown system, which shuts down the engine in an orderly sequence. The number of sensors per engine will depend on the replacement cost of the equipment and the importance of the engine to the system operation.

**CAUTION:** *Intentional overriding of shutdown devices can be dangerous and should never be implemented except under the direction of supervisory personnel.*

Preventive maintenance should be performed on a regular basis. At intervals of about 1 000 hours of operation, power cylinders should be balanced and air/fuel control systems calibrated. The balancing of power

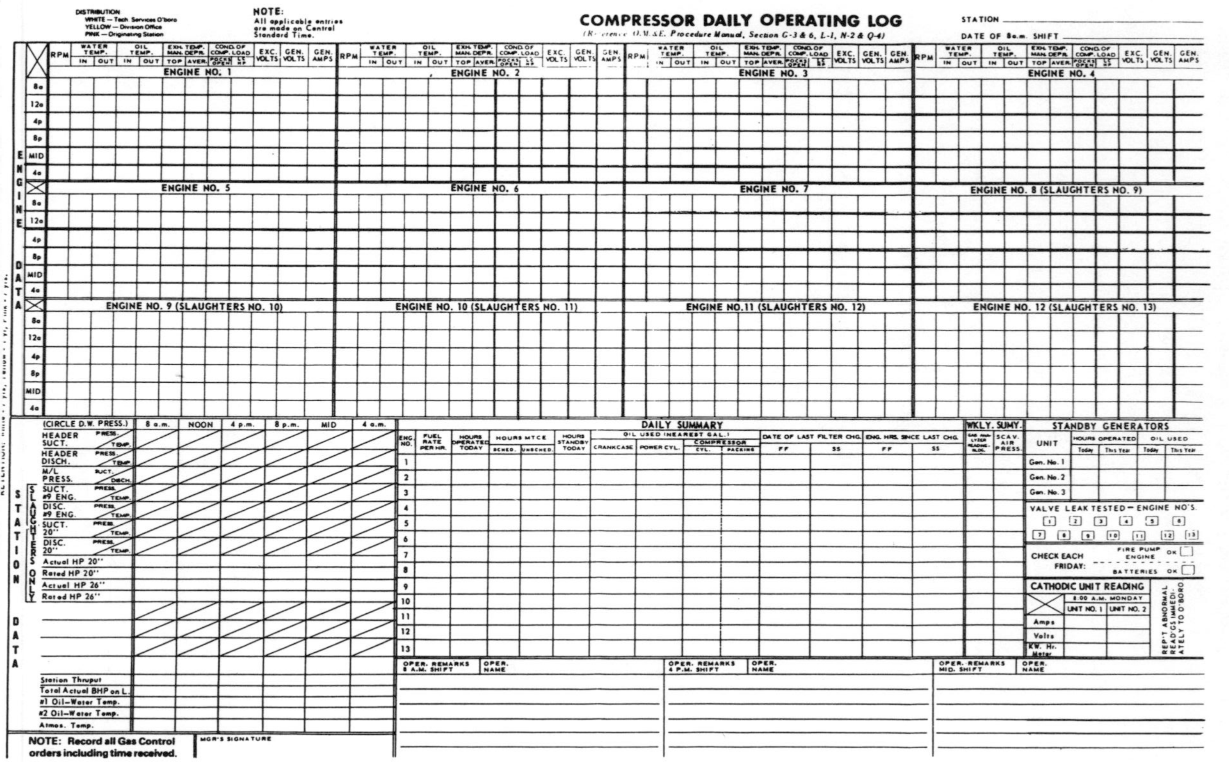

*Figure 69.* Compressor daily operating log

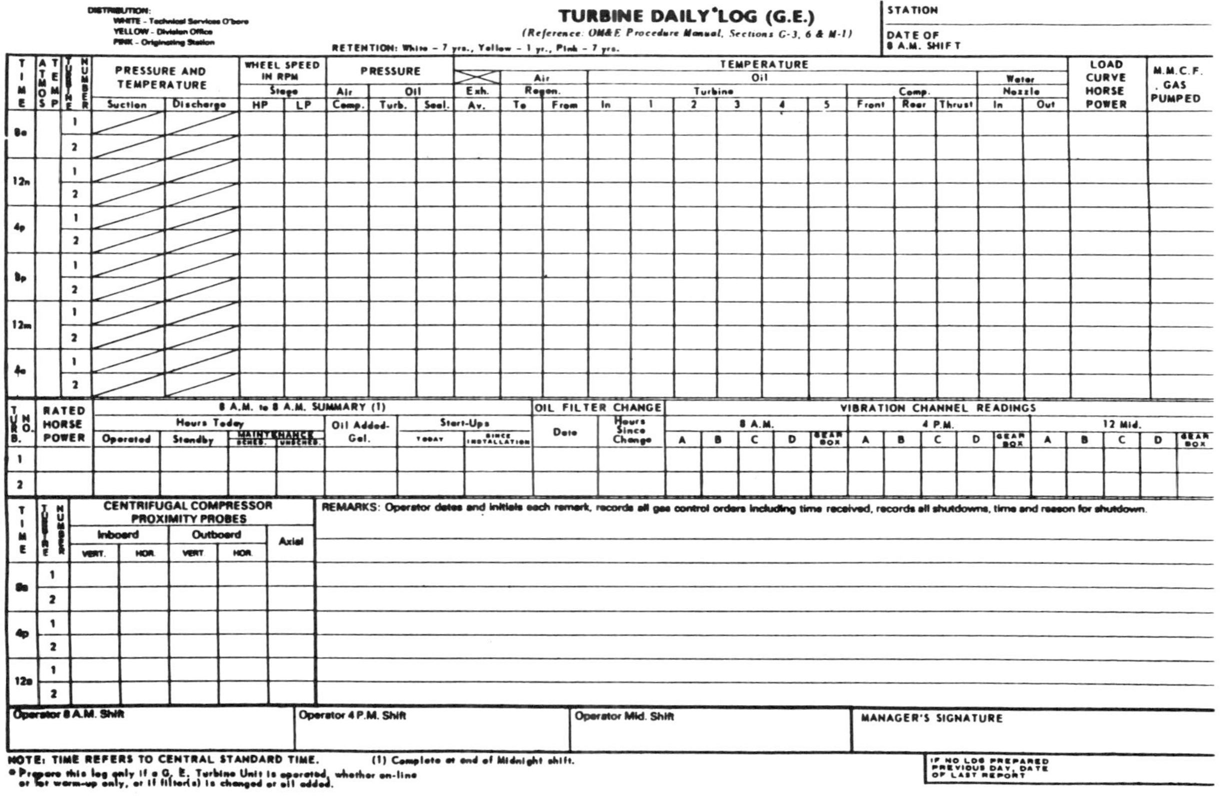


*Figure 70.* Turbine daily log

TG-846 R'82

DISTRIBUTION:
WHITE – Technical Services O'boro
YELLOW – Division Office
PINK – Originating Station

RETENTION: White- 7 yrs, Yellow - 1 yr, Pink - 7 yrs.

## GAS GENERATOR DAILY* LOG (P&W)
*(Reference: OM&E Procedure Manual, Section G-3, 6 & M-1)*

STATION __________  DATE OF 8 a.m. SHIFT 2-4-84

| TIME | ATMOS. TEMP. | TURB NO. | SUCTION | DISCHARGE | N/1 | N/2 | N/3 | A | B | C | D | E | EX-HAUST TEMP. (AVER.) |
|---|---|---|---|---|---|---|---|---|---|---|---|---|---|
| 8a | 34 | 1 | 666  5 | 840  88 | 6170 | 9180 | 4990 | 1.6 | .8 | .4 | 0 | .2 | 900 |
|  |  | 2 |  |  |  |  |  |  |  |  |  |  |  |
| 12n | 47 | 1 | 654  54 | 833  89 | 6180 | 9210 | 5020 | 1.6 | .8 | .4 | 0 | .2 | 920 |
|  |  | 2 |  |  |  |  |  |  |  |  |  |  |  |
| 4p | 46 | 1 | 647  54 | 830  88 | 6130 | 9170 | 5040 | 1.8 | 1.1 | .3 | 0 | .3 | 880 |
|  |  | 2 |  |  |  |  |  |  |  |  |  |  |  |
| 8p | 40 | 1 | 644  54 | 828  88 | 6120 | 9150 | 5060 | 1.8 | 1.0 | .3 | 0 | .3 | 900 |
|  |  | 2 |  |  |  |  |  |  |  |  |  |  |  |
| 12m | 36 | 1 | 644  54 | 818  87 | 6110 | 9140 | 5010 | 1.6 | 1.0 | .4 | 0 | .2 | 885 |
|  |  | 2 |  |  |  |  |  |  |  |  |  |  |  |
| 4a | 36 | 1 | 648  54 | 825  88 | 6170 | 9170 | 5050 | 1.6 | .8 | .4 | 0 | .2 | 895 |
|  |  | 2 |  |  |  |  |  |  |  |  |  |  |  |

Column group headings: PRESSURE AND TEMPERATURE (SUCTION, DISCHARGE); SPEED IN RPM (N/1, N/2, N/3); VIBRATION CHANNELS (A, B, C, D, E).

| TIME | TURB NO. | PS-4 | PS-3 | PT-7 | BREATH | SEAL AIR | BAROM-ETER | FUEL MANI-FOLD PRESS. | GG | SCAV. PUMP | STRAIN-ER IN | LUBE R.T. | SEAL | SEAL OIL DIFF. | E.P.R. (ENG. PRESS. RATIO) | LOAD CURVE HORSE-POWER | M.M.C.F. GAS PUMPED |
|---|---|---|---|---|---|---|---|---|---|---|---|---|---|---|---|---|---|
| 8a | 1 | 144 | 34.0 | 18.8 | 0 | 4.2 | 29.8 | 220 | 44 | 28 | 62 | 30 | 800 | 19 | 2.27 / 1348 | 13800 | 1240 |
| 12n | 1 | 143 | 33.5 | 18.6 | 0 | 4.3 | 29.8 | 220 | 44 | 28 | 63 | 30 | 800 | 18.5 | 2.25 / 1370 | 13700 | 1170 |
| 4p | 1 | 139 | 33.0 | 18.0 | 0 | 5.0 | 29.68 | 210 | 44 | 27 | 63 | 31 | 780 | 18.5 | 2.22 / 1246 | 13200 | 1100 |
| 8p | 1 | 140 | 33.0 | 18.2 | 0 | 4.9 | 29.71 | 212 | 44 | 28 | 63 | 31 | 790 | 19.0 | 2.23 / 1229 | 13300 | 1090 |
| 12m | 1 | 141 | 33.2 | 18.4 | 0 | 5.4 | 29.71 | 214 | 44 | 27 | 63 | 30 | 785 | 19.5 | 2.24 / 1280 | 13400 | 1150 |
| 4a | 1 | 143 | 33.6 | 18.8 | 0 | 5.4 | 29.71 | 217 | 44 | 27 | 63 | 30 | 790 | 19.5 | 2.47 / 1342 | 14000 | 1200 |

Column group headings: AIR PRESSURES (PS-4, PS-3, PT-7, BREATH, SEAL AIR, BAROMETER); OIL PRESSURES (GG, SCAV. PUMP, STRAINER IN, LUBE R.T., SEAL).

### OIL TEMPERATURES

| TIME | TURBINE NUMBER | GAS GENERATOR IN | GAS GENERATOR OUT | REACTION TURBINE IN (LUBE) | REACTION TURBINE OUT (TURB. BRG) | COMPRESSOR THRUST BRG. IN | COMPRESSOR THRUST BRG. OUT | DRIVE END BRG. | IMP. BRG. | SEAL OIL |
|---|---|---|---|---|---|---|---|---|---|---|
| 8a | 1 | 186 | 260 | 114 | 142 | 114 | 133 | 131 | 141 | 119 |
|  | 2 |  |  |  |  |  |  |  |  |  |
| 12n | 1 | 188 | 262 | 115 | 144 | 115 | 136 | 142 | 142 | 120 |
|  | 2 |  |  |  |  |  |  |  |  |  |
| 4p | 1 | 186 | 256 | 116 | 146 | 116 | 136 | 133 | 142 | 120 |
|  | 2 |  |  |  |  |  |  |  |  |  |
| 8p | 1 | 186 | 258 | 115 | 144 | 115 | 136 | 132 | 142 | 120 |
|  | 2 |  |  |  |  |  |  |  |  |  |
| 12m | 1 | 186 | 255 | 115 | 144 | 115 | 135 | 131 | 142 | 119 |
|  | 2 |  |  |  |  |  |  |  |  |  |
| 4a | 1 | 186 | 257 | 114 | 144 | 114 | 136 | 131 | 142 | 118 |
|  | 2 |  |  |  |  |  |  |  |  |  |

### 8 a.m. to 8 a.m. SUMMARY (1)

| TRB NO. | SERIAL NUMBER | RATED HORSE POWER | HOURS TODAY OPERATED | STANDBY | MAINTENANCE SCHED. | MAINTENANCE UNSCHED. | OIL ADDED - GAL. G.G. | OIL ADDED - GAL. R.T./R.C. |
|---|---|---|---|---|---|---|---|---|
| 1 | 634952 | 12000 | 24 | 0 | 0 | 0 | 0 | 0 |
| 2 |  |  |  |  |  |  |  |  |

### CENTRIFUGAL COMP. PROXIMITY PROBES

| TIME | TURB NO. | VERT. | HOR. | AXIAL |
|---|---|---|---|---|
| 8a | 1 | .4 | .4 | +.015 |
|  | 2 |  |  |  |
| 4p | 1 | .4 | .4 | +.08 |
|  | 2 |  |  |  |
| 12a | 1 | .4 | .4 | .11 |
|  | 2 |  |  |  |

REMARKS (Operator dates and initials each remark. Record all gas control orders including time received):

### OIL FILTER CHANGE (1)

| TRB NO. | TYPE | DATE | HOURS SINCE LAST CHANGE |
|---|---|---|---|
| 1 | GG | 1-23-84 |  |
|  | FF-1 | 1-23-84 |  |
|  | FF-2 |  |  |
|  | SS-1 | 1-9-84 |  |
|  | SS-2 |  |  |
| 2 | GG |  |  |
|  | FF-1 |  |  |
|  | FF-2 |  |  |
|  | SS-1 |  |  |
|  | SS-2 |  |  |

DATE OF LAST CARBO-BLAST
AMT. OF CARBO-BLAST USED
DATE OF LAST WATER-WASH

| Operator 8 A.M. Shift | Operator 4 P.M. Shift | Operator Mid. Shift | MANAGER'S SIGNATURE |
|---|---|---|---|
|  |  |  |  |

IF NO LOG PREPARED PREVIOUS DAY, DATE OF LAST REPORT

NOTE: TIME REFERS TO CENTRAL STANDARD TIME    (1) Complete at end of Midnight shift.

* PREPARE THIS LOG ONLY IF A P & W TURBINE UNIT IS OPERATED, WHETHER ON-LINE OR FOR WARM-UP ONLY, OR IF FILTER(S) IS CHANGED OR OIL ADDED.

*Figure 71.* Gas generator daily log

cylinders is done with the aid of a pressure-indicating instrument in an effort to achieve relatively equal firing pressure, or mean effective pressure, in all power cylinders. Adjustments are made in individual fuel control valves, thereby regulating the amount of fuel that is injected into the cylinder. A well-balanced engine would have approximately equal cylinder combustion pressures in all cylinders; consequently, the power output from each cylinder would be essentially equal. Combustion air systems should be inspected to see that these systems are operating efficiently. When the efficiency of an auxiliary air compressor system falls below an acceptable level, corrective action should be taken to restore compressor efficiency.

Major overhaul on a reciprocating engine should be performed when trends established by the monitoring program indicate abnormal operation. A major overhaul would include inspecting all power pistons, power valves, bearings, combustion air systems, and accessory drive equipment. Parts should be replaced as needed, and all protective devices should be calibrated.

## Gas Turbines

A carefully coordinated series of gas turbines used as prime movers are industrial turbines and aircraft-derivative turbines. To properly maintain turbines, operating data recommended by the turbine manufacturer and data required by management should be logged daily, or more frequently, to establish operating trends which will allow maintenance personnel to predict maintenance problems. Operating standards should be available to ensure that gas turbines are operated within company-established parameters.

Maintenance requirements for aircraft-derivative and industrial turbines are described in detail in the following paragraphs.

### Aircraft Derivatives

Base line performance tests should be made on a gas turbine after each major overhaul to establish base line performance data. The performance test data shows turbine wheel speeds at different pressures, corrected exhaust gas temperature, compression ratio across the axial flow compressor, and fuel rates. A typical performance map is shown in Figure 72.

Three times each week performance data should be logged and calculations made to compare actual engine performance against the base line performance curve. A typical performance calculation data sheet is shown in Figure 73. A comparison of the two curves will in-

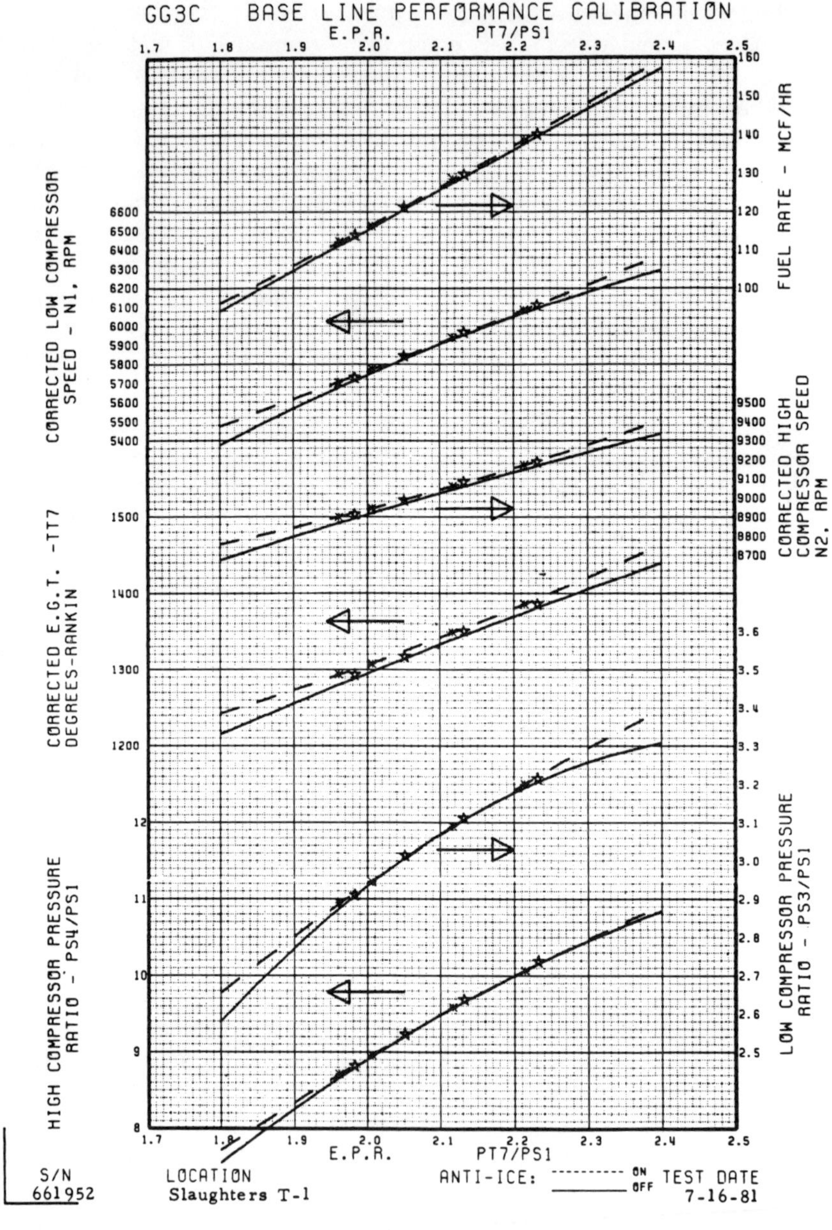


*Figure 72.* **Base line performance calibration**

## P&W GG3C-4 PERFORMANCE CALCULATIONS

NOTE: SHADED AREAS INDICATE WHERE READINGS (OBSERVED DATA) ARE TO BE ENTERED. UNSHADED AREAS ARE FOR CALCULATED VALUES.

| LOCATION | | | TURBINE SERIAL NUMBER | TIME | DATE |
|---|---|---|---|---|---|
| Hardinsburg, Ky. | T-2 | | 634059 | 8:40 | 1-21-83 |

| ITEM NO. | DESCRIPTION | CALCULATION FORMULA, IF APPLICABLE— CIRCLED NO'S REFER TO ITEM NO. | CURRENT VALUE | BASE LINE CURVE VALUE | DIFFERENCE FROM BASE | LAST REPORT'S DIFF. FROM BASE |
|---|---|---|---|---|---|---|
| 1 | Tt2 Observed °F Plenum Chamber | | 31° | | | |
| 2 | N1 Observed RPM | | 5970 | | | |
| 3 | N1 Corrected RPM | ② ÷ √θ | 6137 | 6140 | +3 | +9 |
| 4 | N2 Observed RPM | | 9060 | | | |
| 5 | N2 Corrected RPM | ④ ÷ √θ | 9314 | 9320 | +6 | +57 |
| 6 | Tt7 Observed °F (see Area A calculation) | | 847 | | | |
| 7 | Tt7 Corrected °R | ( ⑥ + 460 ) ÷ θ | 1381 | 1395 | | |
| 8 | Barometric Pressure " Hg | | 29.30 | RECORD INDIVIDUAL Tt7 OBSERVED TEMPERATURES BELOW AND ENTER AVERAGE AS ITEM NO. 6 | | |
| 9 | Barometric Pressure PSIA | ⑧ × 0.491 | 14.38 | | | |
| 10 | Pt7 Observed " Hg (Clarksdale Only) | | | Tt7-1 840 | Ⓐ | |
| 11 | Pt7 Observed PSIG | | 17.6 | Tt7-2 845 | | |
| 12 | Pt7 Manometer Temperature °F | | 74° | Tt7-3 848 | | |
| 13 | Pt7 Manometer Correction PSIG (from curves) | | .07 | Tt7-4 832 | | |
| 14 | Pt7 Corrected PSIG | ⑪ − ⑬ | 17.53 | Tt7-5 792 | | |
| 15 | Pt7 Absolute PSIA | ⑭ + ⑨ | 31.91 | Tt7-6 895 | | |
| 16 | E. P. R. Pt7 /Ps1 | ⑮ ÷ ⑨ | 2.21 | Tt7-7 875 | | |
| 17 | Ps3 Observed PSIG | | 32.8 | Tt7-8 845 | | |
| 18 | Ps3 Absolute PSIA | ⑰ + ⑨ | 47.18 | TOTAL 6772 | ÷ 8 = 847 | |
| 19 | Ps3 /Ps1 | ⑱ ÷ ⑨ | 3.28 | 3.25 | −.03 | −.02 |
| 20 | Ps4 Observed PSIG | | 140 | | | |
| 21 | Ps4 Absolute PSIA | ⑳ + ⑨ | 154.38 | | | |
| 22 | Ps4 /Ps1 | ㉑ ÷ ⑨ | 10.73 | 10.55 | −.18 | −.11 |
| 23 | Fuel Differential " H₂O | | 51 | | | |
| 24 | Fuel Pressure PSIA | | 660 | | | |
| 25 | Fuel Temperature °F | | 69° | | | |
| 26 | Fuel Rate MCF / HR (from tables) | | 133.8 | 136.5 | +2.7 | +4.6 |
| 27 | Anti-Ice System | | ☐ OFF ☒ ON | | | |
| 28 | | | | | | |
| 29 | | | | | | |
| 30 | | | | | | |

| DATE OF LAST CARBON BLAST | 1-21-83 |
|---|---|
| AMOUNT OF CARBON BLAST USED | 25 # |
| DATE OF LAST WATER - WASH | 1-7-83 |

REMARKS

| PREPARED BY | MANAGER'S SIGNATURE |
|---|---|



*Figure 73.* **Turbine performance calculations**

dicate which components of the turbine require maintenance and whether turbine failure is imminent.

A performance trend analysis is shown in Figure 74. All vibration sensors permanently mounted on the turbine should be checked weekly with a portable vibration analyzer to verify the vibration levels shown by the monitor.

Hot section inspections on aircraft-derivative gas turbines are performed after a fixed number of operating hours as recommended by the engine manufacturer, or when condition monitoring indicates problems in the hot section area. A hot section consists of the fuel injection system, combustion chambers, and exhaust gas passages. Hot section repair kits should be available on location and contain sufficient parts to allow maintenance personnel to replace damaged parts found during the inspection.

The gas turbine should be removed and shipped to a reputable overhaul and maintenance facility when major maintenance is required. Sufficient spare gas turbines should be maintained in the company's fleet to replace a turbine removed for overhaul. The number of spare turbines required will depend upon the use factor, the number of turbines in the fleet, the operating history, and the turnaround time in an overhaul and maintenance shop.

### Industrial Turbines

Industrial gas turbine test data should be taken once a week so that turbine performance can be calculated. A typical data sheet is shown in Figure 75. When the adiabatic efficiency of the axial flow compressor drops to a specified, predetermined level, a cleaning agent—such as nutshells or inorganic compounds—should be injected into the inlet air system to clean the air compressor and restore adiabatic efficiency to an acceptable level. The amount of cleaning agent injected should be determined from operating data. When injected solids will no longer restore compressor efficiency to an acceptable level, a water wash should be applied to restore compressor efficiency. Operating standards should be available to describe procedures for cleaning turbines on-stream—with both solid particles and liquid wash. This monitoring of the axial flow compressor efficiency shall apply to both regenerator and nonregenerating turbines.

Data should be accumulated on a weekly basis for industrial turbines with a regenerative cycle, and the regenerator effectiveness should be calculated. This is illustrated on the data sheet shown in Figure 75. These calculations show not only the loss of regenerator effectiveness but also the output power loss. A performance evaluation can be made

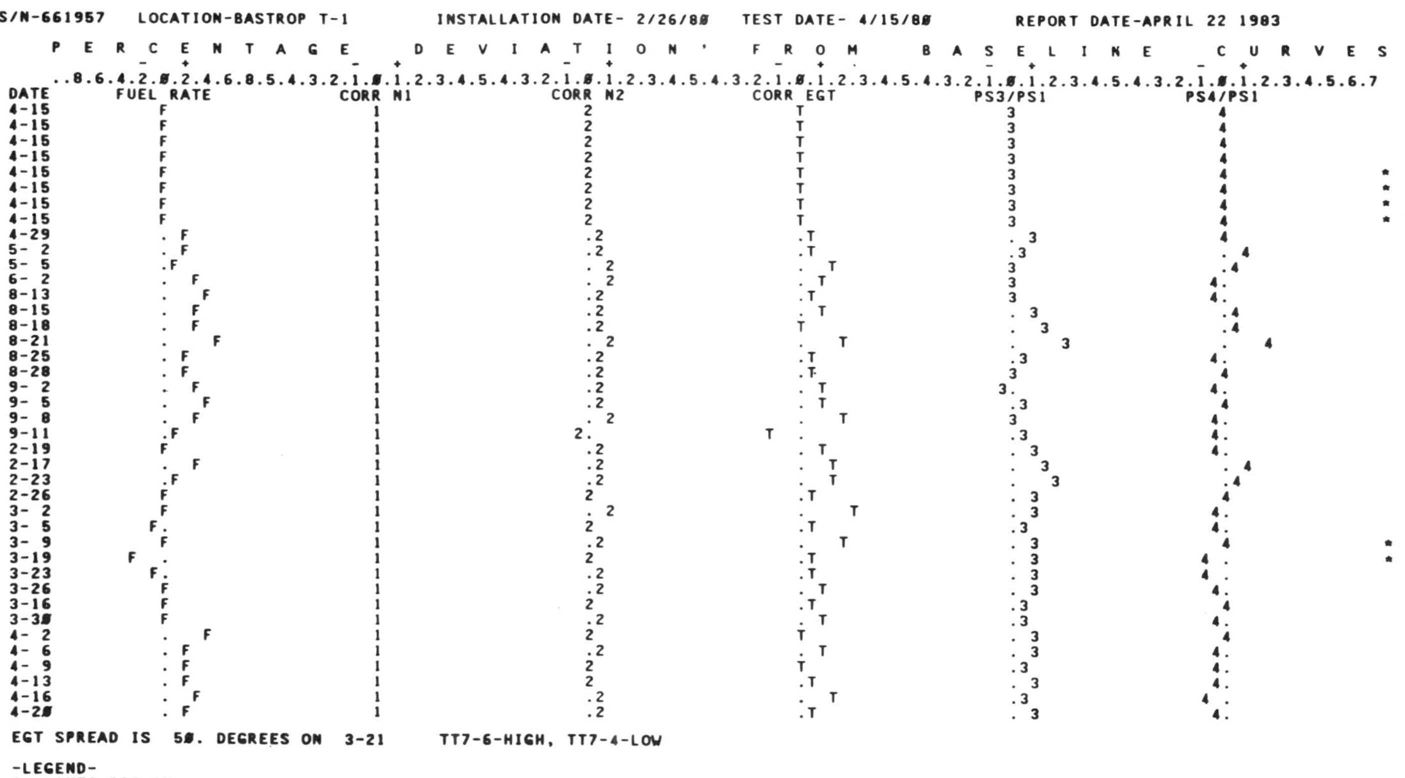



*Figure 74.* Condition monitoring performance trends

## G.E. GAS TURBINE PERFORMANCE CALCULATIONS (PREPARE WEEKLY) TG-905 R '84

NOTE: SHADED AREAS INDICATE WHERE READINGS (OBSERVED DATA) ARE TO BE ENTERED. UNSHADED AREAS ARE FOR CALCULATED VALUES.

| LOCATION: Lake Cormorant, Ms | PREPARED BY | TIME: 7:00 AM | DATE OF TEST: 10-17-84 |
|---|---|---|---|

| ITEM NO. | DESCRIPTION (CALCULATION FORMULA, IF APPLICABLE— CIRCLED NOS. REFER TO ITEM NO.) | Turbine # 3 LEFT | Turbine # 3 RIGHT | Turbine # 4 LEFT | Turbine # 4 RIGHT |
|---|---|---|---|---|---|
| | **TURBINE REGENERATOR PERFORMANCE** | | | | |
| 1 | Air Compressor Discharge Temperature °F | 463 | 463 | 515 | 515 |
| 2 | Regenerator Air Inlet Temperature °F | 565 | 580 | 515 | 515 |
| 3 | Regenerator Air Outlet Temperature °F | 861 | 866 | 949 | 940 |
| 4 | Regenerator Exhaust Gas Inlet Temperature °F (See Note) | 985 | 985 | 998 | 998 |
| 5 | Temperature Efficiency $\frac{③-②}{④-②} \times 100$ | 70.48 % | 68.10 % | 89.85 % | 87.99 % |
| 6 | Barometric Pressure "Hg | 29.94 | | 29.74 | |
| 7 | Barometric Pressure PSIA ⑥ × .491 | 14.70 | | 14.61 | |
| 8 | Air Inlet Pressure PSIG | 74.5 | | 79 | |
| 9 | Air Inlet Pressure PSIA ⑧ + ⑦ | 89.20 | | 93.61 | |
| 10 | Air Pressure Drop - ▲ P PSI | 2.16 | 1.96 | .884 | .835 |
| 11 | Air Pressure Drop Percent ⑩ ÷ ⑨ × 100 | 2.42 % | 2.19 % | .944 % | .892 % |
| 12 | Exhaust Gas Inlet Pressure "$H_2O$ | 10.0 | | 8.75 | |
| 13 | Exhaust Gas Inlet Pressure PSIG ⑫ × .036 | .360 | | .315 | |
| 14 | Exhaust Gas Inlet Pressure PSIA ⑬ + ⑦ | 15.06 | | 14.93 | |
| 15 | Exhaust Gas Pressure Drop % ⑬ ÷ ⑭ × 100 | 2.39 % | 2.39 % | 2.11 % | 2.11 % |
| 16 | Total Pressure Drop Percent ⑪ + ⑮ | 4.81 % | 4.58 % | 3.05 % | 3.00 % |

| ITEM NO. | DESCRIPTION | Turbine # 3 | EXHAUST TEMPERATURES | | Turbine # 4 | EXHAUST TEMPERATURES | |
|---|---|---|---|---|---|---|---|
| | **TURBINE AIR COMPRESSOR PERFORMANCE** | | | | | | |
| 17 | Inlet Temperature (Plenum Chamber) °F | 59.8 | 1 | 982 | 79.0 | 1 | 1001 |
| 18 | Plenum Chamber Pressure Drop "$H_2O$ | 3.6 | 2 | — | 4.0 | 2 | 1008 |
| 19 | Plenum Chamber Pressure Drop PSI ⑱ × .036 | .130 | 3 | 985 | .144 | 3 | 1010 |
| 20 | Inlet Pressure PSIA ⑦ − ⑲ | 14.57 | 4 | 992 | 14.47 | 4 | 999 |
| 21 | Compression Ratio ⑨ ÷ ⑳ | 6.12 | 5 | 983 | 6.47 | 5 | 1009 |
| 22 | Compressor Efficiency (from curves) ㉑, USE ⑰, ① | 86 | 6 | — | 89.5 | 6 | 1011 |
| 23 | H. P. Turbine Speed RPM | 6920 | 7 | 985 | 6915 | 7 | 1012 |
| | **CENTRIFUGAL COMPRESSOR DATA** | Turbine # 3 | 8 | — | Turbine # 4 | 8 | 1008 |
| 24 | LP Turbine Speed RPM | 5710 | 9 | 977 | 5822 | 9 | 1001 |
| 25 | Suction Nozzle Pressure (Nearest 0.1 PSIG) | 595 | 10 | — | 719 | 10 | 978 |
| 26 | Discharge Nozzle Pressure (Nearest 0.1 PSIG) | 809 | 11 | — | 842 | 11 | 970 |
| 27 | Suction/Discharge Temperature (Nearest 0.1 °F) | 70.4 / 118.3 | 12 | 993 | 75.2 / 100.6 | 12 | 979 |
| 28 | Flow Pressure PSIA | 620 | Avg. | 985 | 740 | Avg. | 998 |
| 29 | Flow Differential "$H_2O$ | 27.5 | | | 76.0 | | Torque Meter Horsepower (Clarksdale only) |
| 30 | Fuel Pressure PSIA | 260 | NOTE: If regenerator is a Harrison, enter average exhaust temperature in Item No. 4. | | 240 | | |
| 31 | Fuel Differential "$H_2O$ /Fuel Temp. °F | 22 / 62.5 | | | 23 / 67 | | |
| 32 | Type of Control - Fuel (F) Nozzle (N) /Nozzle Angle | F / 2.5° | | | F / 0° | | |
| 33 | Load Curve Horsepower/Volume | 8300 / 21.3 | | | 19300 / 50.6 | | |

| | | A | B | C | D | GEAR BOX | A | B | C | D | GEAR BOX |
|---|---|---|---|---|---|---|---|---|---|---|---|
| | Vibration Channel Readings | .15 | .15 | .15 | 05 | .17 | .05 | .14 | .12 | .12 | .02 |

| | | INBOARD VERT | INBOARD HOR | OUTBOARD VERT | OUTBOARD HOR | AXIAL | INBOARD VERT | INBOARD HOR | OUTBOARD VERT | OUTBOARD HOR | AXIAL |
|---|---|---|---|---|---|---|---|---|---|---|---|
| | Centrifugal Compressor Proximity Probes | N/A | N/A | N/A | N/A | +9.5 | 1.4 | 1.8 | 28 | .2 | +3 |

| | Date Turbine Last Cleaned / Amount of Compound Used | 10-17-84 / 3 lbs. | 10-22-84 / 3 lbs |
|---|---|---|---|

DISTRIBUTION: White - Tech. Services, Owensboro; Yellow - Division Office (Division I) Area Office (Division II & III); Pink - Originating Station
RETENTION: Until next major overhaul.

***Figure 75.*** **Gas turbine performance calculations**

from the data recorded in the field. Typical performance data is shown in Figure 76.

At some level of ineffectiveness it will become necessary to repair the regenerator. Historical operating and performance data will show when the repair is economical. Operating standards should be available for station personnel for testing and repairing regenerators.

Hot section inspections on industrial gas turbines normally are performed on an annual basis or after 8 000 hours of operation. These inspections include visually inspecting the combustion chambers and the components associated with gas combustion. The hot gas path component should be inspected with a borescope to determine if any erosion has occurred, to look for cracked or damaged inlet guide vanes, and to identify damaged turbine components. All control systems should be calibrated during the annual inspection. Operating standards should be available to describe how to perform an 8 000-hour inspection. A detailed repair report should be made of the inspection.

A major inspection is performed after 30 000 to 50 000 hours of operation. This involves removing the axial flow compressor, the turbine section, and the combustion section. Components that need to be repaired or investigated should be sent to a reputable overhaul and maintenance facility. All control systems should be calibrated. A detailed report should be made of the engine condition and the required repairs.


## MAINTAINING COMPRESSORS

Gas compressors used at compressor stations to pump gas can be divided into two major categories: reciprocating compressors and centrifugal compressors.


### Reciprocating Compressors

To determine the condition of a reciprocating compressor, operating data should be logged daily so that the adiabatic efficiency of the compressor can be calculated. The compressors should be checked daily with an ultrasonic leak detector to locate leaking compressor valves. If the compressor is operating with a high use factor, a minimum level of adiabatic efficiency should be established to determine when a compressor valve should be changed. Leaking compressor valves should be replaced when the compressor efficiency falls 2 percentage points below the established norm. If a compressor is operating with a high use factor and no spare equipment is available, compressor valves should be inspected when the compressor efficiency falls 3 to 5 percent below the norm. If replacement of the compressor valve will not restore the effi-

GENERAL ELECTRIC TURBINE PERFORMANCE RECORD

CLARKSDALE T-3

REPORT DATE    8-23-83                    NEMA RATING   12090 HP                    



| DATE | AIR COMPRESSOR INLET TEMP | DIS PRES | % DEV* | ADIA EFF | LP EXH PRES H2O | REGEN % PRES DROP L | REGEN % PRES DROP R | EFFECTIVENESS L | EFFECTIVENESS R | EXHAUST TEMP AVERAGE ACTUAL | EXHAUST TEMP DESIGN | CENTRIFUGAL BHP ACTUAL | BHP % DEV | BHP DESIGN* | RATED SPEED LP | RATED SPEED HP | FUEL RATE ACTUAL | FUEL RATE % DEV | FUEL RATE DESIGN* |
|---|---|---|---|---|---|---|---|---|---|---|---|---|---|---|---|---|---|---|---|
| --------CLARKSDALE T-3 NEW INSTALLATION-------------------- | | | | | | | | | | | | | | | | | | | |
| 1-12-83 | 35.0 | 88.0 | 0.0 | ***** | 9.1 | 0.0 | 0.0 | 0.0 | 0.0 | 1007 | 984 | 13312 | 0.0 | 0 | 100. | 96.4 | 7.91 | 0.0 | 0.00 |
| 1-27-83 | 38.0 | 88.5 | 0.0 | ***** | 8.7 | 0.0 | 0.0 | 0.0 | 0.0 | 1007 | 987 | 12930 | 0.0 | 0 | 99. | 98.0 | 8.55 | 0.0 | 0.00 |
| -----------PERFORMANCE TEST DATA--------------------------- | | | | | | | | | | | | | | | | | | | |
| 1-27-83 | 35.3 | 87.3 | 0.0 | 76.0 | 8.6 | 3.5 | 3.2 | 88.0 | 88.0 | 1005 | 985 | 12155 | 0.0 | 0 | 99. | 97.9 | 8.48 | 0.0 | 0.00 |
| 1-27-83 | 39.1 | 89.4 | 0.0 | 81.0 | 9.2 | 3.5 | 3.3 | 88.6 | 88.4 | 1007 | 987 | 12645 | -15.2 | 14913 | 99. | 100.0 | 8.36 | 7.3 | 7.79 |
| 2- 1-83 | 56.3 | 83.7 | 0.0 | 81.7 | 8.4 | 3.1 | 3.2 | 88.6 | 87.8 | 1012 | 999 | 11126 | -17.3 | 13448 | 95. | 100.0 | 8.80 | 9.4 | 8.05 |
| 2- 1-83 | 58.6 | 83.0 | 0.0 | 81.8 | 8.6 | 3.1 | 3.1 | 88.6 | 89.4 | 1013 | 1001 | 10897 | -18.0 | 13296 | 95. | 100.0 | 8.89 | 10.0 | 8.08 |
| 2- 1-83 | 59.6 | 82.4 | 0.0 | 81.7 | 8.6 | 3.2 | 3.2 | 88.3 | 87.5 | 1013 | 1001 | 10821 | -18.1 | 13212 | 95. | 100.0 | 8.89 | 9.8 | 8.10 |
| 2- 1-83 | 59.1 | 82.5 | 0.0 | 81.5 | 8.4 | 3.1 | 3.2 | 88.3 | 87.5 | 1013 | 1001 | 10878 | -18.1 | 13288 | 95. | 100.0 | 8.84 | 9.4 | 8.08 |
| 2- 1-83 | 59.2 | 81.9 | 0.0 | 81.6 | 8.4 | 3.1 | 3.1 | 88.5 | 87.4 | 1008 | 1001 | 10799 | -18.7 | 13279 | 95. | 99.7 | 8.81 | 9.0 | 8.08 |
| 2- 2-83 | 36.2 | 88.0 | 0.0 | 81.0 | 8.4 | 3.2 | 3.2 | 88.3 | 87.2 | 1009 | 985 | 12224 | -16.7 | 14669 | 99. | 100.0 | 8.49 | 9.4 | 7.76 |
| 2- 2-83 | 35.8 | 80.2 | 0.0 | 80.9 | 8.9 | 3.3 | 3.3 | 88.5 | 87.4 | 1009 | 985 | 12315 | -16.2 | 14698 | 99. | 100.0 | 8.51 | 9.7 | 7.76 |
| 2- 2-83 | 36.1 | 89.8 | 0.0 | 85.3 | 9.3 | 3.4 | 3.5 | 88.2 | 87.6 | 1000 | 985 | 13268 | 0.0 | 0 | 101. | 97.1 | 8.05 | 0.0 | 0.00 |
| --------END PERFORMANCE TEST DATA--------------------------- | | | | | | | | | | | | | | | | | | | |
| 2- 2-83 | 36.2 | 88.0 | 0.0 | ***** | 8.4 | 0.0 | 0.0 | 0.0 | 0.0 | 1009 | 985 | 12238 | -18.0 | 14928 | 99. | 100.0 | 8.49 | 9.4 | 7.76 |
| 2-10-83 | 50.0 | 89.0 | 0.0 | ***** | 9.4 | 0.0 | 0.0 | 0.0 | 0.0 | 1007 | 995 | 14197 | -0.6 | 14289 | 100. | 100.0 | 7.48 | -5.6 | 7.92 |
| 2-18-83 | 41.0 | 88.5 | 0.0 | ***** | 9.2 | 0.0 | 0.0 | 0.0 | 0.0 | 1005 | 989 | 13657 | 0.0 | 0 | 99. | 98.2 | 7.72 | 0.0 | 0.00 |
| --------PERFORMANCE TEST DATA STATION EQUIP.------------------ | | | | | | | | | | | | | | | | | | | |
| 5- 3-83 | 68.5 | 88.7 | 0.0 | 86.3 | 9.3 | 3.3 | 3.4 | 87.8 | 86.9 | 996 | 1004 | 13172 | 1.9 | 12928 | 99. | 100.0 | 7.95 | -3.3 | 8.22 |
| 5- 3-83 | 68.6 | 87.9 | 0.0 | 86.5 | 9.3 | 3.3 | 3.4 | 87.9 | 87.1 | 1000 | 1004 | 13318 | 3.1 | 12921 | 99. | 99.4 | 7.76 | -5.6 | 8.22 |
| 5- 4-83 | 58.8 | 91.4 | 0.0 | 86.3 | 9.4 | 3.4 | 3.4 | 87.9 | 86.9 | 1000 | 1001 | 14331 | 4.8 | 13676 | 101. | 100.0 | 7.64 | -4.9 | 8.03 |
| 5- 4-83 | 62.4 | 90.6 | 0.0 | 86.3 | 9.5 | 3.4 | 3.3 | 88.1 | 87.0 | 1001 | 1002 | 13637 | 1.5 | 13437 | 101. | 100.0 | 7.90 | -2.4 | 8.09 |
| 5- 4-83 | 63.9 | 90.7 | 0.0 | 86.3 | 9.5 | 3.4 | 3.4 | 87.6 | 86.8 | 1002 | 1002 | 14323 | 7.4 | 13338 | 101. | 100.0 | 7.50 | -7.7 | 8.12 |
| 5- 4-83 | 66.0 | 90.2 | 0.0 | 86.4 | 9.6 | 3.4 | 3.3 | 87.6 | 86.7 | 1003 | 1003 | 14136 | 7.1 | 13193 | 101. | 100.0 | 7.57 | -7.2 | 8.16 |
| 5- 4-83 | 66.6 | 89.8 | 0.0 | 86.2 | 9.6 | 3.4 | 3.3 | 87.6 | 86.7 | 1003 | 1003 | 13896 | 5.7 | 13147 | 102. | 100.0 | 7.62 | -6.7 | 8.17 |
| 5- 4-83 | 69.4 | 89.2 | 0.0 | 86.6 | 9.6 | 3.4 | 3.3 | 87.6 | 86.5 | 1004 | 1004 | 13844 | 6.9 | 12950 | 103. | 100.0 | 7.58 | -7.8 | 8.22 |
| 5- 5-83 | 64.4 | 91.2 | 0.0 | 86.8 | 9.4 | 3.3 | 3.3 | 87.6 | 86.8 | 1002 | 1003 | 13724 | 3.4 | 13278 | 99. | 100.0 | 7.80 | -4.2 | 8.14 |
| --------PERFORMANCE TEST DATA TECH. SERV. EQUIP.--- | | | | | | | | | | | | | | | | | | | |
| 5- 3-83 | 68.5 | 88.5 | 0.0 | 86.2 | 9.3 | 3.3 | 3.4 | 91.0 | 90.2 | 979 | 1004 | 13172 | 1.9 | 12928 | 99. | 100.0 | 7.95 | -3.3 | 8.22 |
| 5- 3-83 | 68.6 | 87.9 | 0.0 | 86.5 | 9.3 | 3.3 | 3.4 | 90.8 | 89.9 | 985 | 1004 | 13318 | 3.1 | 12921 | 99. | 99.4 | 7.76 | -5.6 | 8.22 |
| 5- 4-83 | 58.8 | 91.0 | 0.0 | 86.1 | 9.4 | 3.4 | 3.4 | 90.9 | 89.9 | 984 | 1001 | 14331 | 4.8 | 13676 | 101. | 100.0 | 7.64 | -4.9 | 8.03 |

*CALCULATED AT RATED HP SPEED ONLY

*Figure 76.* Turbine performance record

ciency to an acceptable level, the compressor piston should be removed, and the compressor cylinder and rings should be inspected. A detailed repair report should be prepared when a piston is removed. Compressor rod packing should be replaced or inspected only when there is an indication of gas blow-by. Compressor rod packing replacement should be done in accordance with the manufacturer's recommended guidelines.

Overhauling reciprocating compressors on routine intervals is not required if regular tests are made to ensure that the compressor is functioning properly.

Vibration sensors should be located on each compressor, and should be sensitive enough to trip the engine shutdown system in case of a loose piston rod, a loose piston nut, or any other unusual vibration problem occurring in the compressor.

## Centrifugal Compressors

The routine maintenance of centrifugal compressors involves primarily observing data and conducting routine performance checks. Operating data that includes seal oil and lube oil pressures, vibration levels, and oil consumption will establish trends that indicate potential problems in the lube oil or seal oil system. Excessive seal oil consumption or pressure loss indicates possible problems that may require dismantling of the compressor. This data should be recorded in the daily operating log. An increase in vibration levels also will indicate a need to dismantle the compressor to check for leaking seals, worn or damaged rotor shaft or impellers, or rotor imbalance.

Performance data should be recorded so that the compressor efficiency can be calculated. By measuring the temperature rise and pressure rise across the compressor, compressor adiabatic efficiency can be determined. By measuring the volume pumped by the compressor and compressor speed, a $Q/N$ value can be calculated, where $Q$ is the volume of gas and $N$ is the compressor speed. A compressor efficiency versus $Q/N$ value curve can be plotted and compared to the original performance curve. When a trend is established that shows compressor efficiency has fallen 3 to 4 percent below the original performance curve, an attempt should be made to clean the compressor on-stream by injecting a water wash solution. An operating standard should be available to describe the on-stream wash process. Continuous-run times of 24 000 hours between washes are feasible.

If the wash process fails to restore the compressor efficiency to an acceptable level, the compressor should be dismantled and inspected for internal leaks, damaged rotor impellers, or damage to other components that would cause loss of efficiency.

If operating data shows no loss of performance or problems in the seal system, a routine compressor overhaul is not required. In general a centrifugal compressor only needs to be overhauled when operating data indicates problems.

## MAINTAINING AUXILIARY SYSTEMS

Auxiliary equipment at a compressor station usually includes support equipment for the compressor station such as electrical power generating units or standby electrical power generator units, air compressors, fuel gas heating systems, and boilers.

### Electric Power Generated by Reciprocating Engines

At compressor stations where all electrical power is produced by reciprocating engines driving generators, a maintenance program for these units should follow the same guidelines established for reciprocating engines driving gas compressors. When more than one generator unit is required to meet the load demand, the load carried by each unit should be such that there will be sufficient reserve horsepower available to carry the load should a single unit fall off line. An engine rotation system should be employed to balance the operating hours on each unit.

The generators on these units should be inspected as described in the electrical maintenance program usually supplied by the manufacturer.

### Air Compressors

The same maintenance guidelines should apply for reciprocating engines that drive air compressors as those established for other reciprocating engines at the station. When electric motors are used to drive the air compressors, the same maintenance guidelines established for other station electric motors should be applied.

Once each week the suction and discharge valves in the air compressor should be inspected with an ultrasonic leak detector or with an engine analyzer. Leaking valves should be replaced. The intermediate pressure and final discharge pressure should be observed weekly. When the range between these pressures exceeds predetermined limits, the compressor should be overhauled. The overhaul would include close examination of the pistons and the valves, and replacing excessively worn parts.

## Boilers

Boilers that provide hot water for gas heaters and building heat should be checked monthly. To prevent scaling in the boilers, the pH of the boiler water should be checked regularly and maintained at the manufacturer's recommended level. The boiler control should be calibrated semiannually.

## MAINTAINING PIPING AND VESSELS

High pressure piping at a compressor station presents a potentially hazardous situation if proper maintenance practices are not established.

At least daily immediately after start-up and on a weekly basis thereafter, the station maintenance specialist or maintenance personnel should inspect all pipe supports and restraining straps or blocks, and look for any unusual vibration in the piping. Experience will indicate critical points in the high pressure piping supports that require close surveillance. All high pressure piping should be restrained so that movement is minimal.

All high pressure yard piping should be inspected annually for corrosion. At least once every 3 years high pressure restraining straps should be removed, and the piping underneath the straps should be inspected for corrosion. Piping that is corroded excessively should be repaired or replaced as required.

High pressure piping encapsulated for noise abatement purposes should be inspected routinely to ensure that moisture is not seeping under the encapsulating material. A semiannual ultrasonic inspection should be performed to detect leaks.

**NOTE:** *Personnel should be trained in and follow manufacturer instructions concerning torquing on all gas piping and related equipment.*

## MAINTENANCE OF ELECTRICAL EQUIPMENT

The electrical maintenance requirements at a compressor station vary with the design and age of the facility. An electrical maintenance program should include switchgear, motors, ac and dc generators, exciters, voltage regulators, gauges, terminals, and cables. Components in monitoring systems, control systems, and sequencing systems should be maintained in accordance with the suggested maintenance program for those systems. A master file should be maintained that contains the technical data for each electrical apparatus: type of inspection required, inspection dates, repairs, and maintenance performed. An excellent reference book on details of electrical equipment maintenance is *Westinghouse Electrical Hints* published by Westinghouse Electric Corporation.

## AC Generators, Alternator Type

At stations where electrical power is generated on-site, the station operator should log on a daily basis the generator voltage, output frequency, ac current in amperes, dc current in amperes, and ac power in kilowatts. This data will provide a useful performance and a power output record.

A maintenance program for generators will include daily observation by the operator and quarterly inspection by the electrician. On a daily basis the operator should look for drifts in voltages and frequency, and arcing at collector rings. On a quarterly schedule the generator should be checked for etched or grooved collector rings, visible sparking at brushes, worn exciter windings and ac windings, wear or binding of brushes, abnormal stator temperatures, and unusual operating noises.

On auxiliary ac generators that are driven by accessory gears of prime movers, the quarterly inspection described for power generators should be followed. Every 3 years generators should be disassembled, and cleaned, and the bearings should be repacked with grease.

## AC and DC Motors Driving Fin Fans, Pumps, Etc.

On a quarterly basis ac and dc motors should be checked for bearing noise, magnetic hum, excessive vibration, high operating temperature, and deteriorated motor lead insulation. In addition, the commutator and brushes on dc motors should be inspected.

On an annual basis the motor should be disassembled and cleaned, and the bearings should be replaced or repacked with grease. The windings should be meggered and checked for deterioration. The commutator should be checked for looseness, wear, and grooving. All wires should be checked for loose or broken connections and cracked or deteriorated insulation.

## MAINTAINING INSTRUMENTS AND CONTROLS

To maintain a reliable and dependable maintenance program for control, monitoring, and sequencing systems at a station, management should have a master tabulation list of all components in each system. This tabulation should show the scheduled test data and operating parameters for each component in the system.

## Electronic Control Systems

All electronic monitoring and control systems should be calibrated at least on an annual basis. Electronic instrumentation subjected to seasonal ambient temperature variations has a tendency to drift, and requires frequent zero and span adjustments. Calibration instruments usually are described in the operating technical standards or in the manufacturer's manual. System components should be maintained in an environment that is free of dust and foreign materials.

## Pneumatic Control Systems

All pneumatic monitoring and control systems should be calibrated on an annual basis. Operating technical standards or the manufacturer's manual usually contain a list of the required calibration instruments. Components that are exposed to outside weather conditions should be enclosed in weatherproof housings. The air supply for pneumatic systems should be processed so that all moisture and foreign materials are removed.

In addition to the annual inspection and calibration, controls specialists or maintenance specialists should perform a weekly inspection. System components should be checked to see that pressures, voltages, and vibration are within a tolerable range and that vents are free from debris. In general the inspector should be looking for anything unusual that ultimately might affect the performance of the system.

## MAINTAINING STRUCTURES AND HOISTING EQUIPMENT

Structures (compressor building, auxiliary building, etc.) will not require a significant amount of attention if painted periodically with a good quality, rust-preventive paint.

Overhead crane and hoisting equipment should be inspected annually, and should meet or exceed all applicable federal regulations and requirements. The inspection should include a thorough examination for wear, broken strands, or kinks on the lifting cables; for slippage in the brake system; and for cracks in the lifting hooks.

# EQUIPMENT TESTING, INSPECTION, AND MONITORING

Compressor station equipment and systems can be maintained at top operating efficiencies by utilizing testing equipment, by implementing comprehensive testing and inspection programs, and by thorough operational monitoring.

## TEST EQUIPMENT

By utilizing testing equipment, skilled operators maintain station equipment at peak performance and identify problems before they lead to catastrophic failures. The equipment used will vary with station design and components, but commonly will include an engine analyzer, a power cylinder load balancer, and optical aids.

### Engine Analyzer

The engine analyzer is an instrument that permits an examination of the operating parameters of each engine cylinder. Using the analyzer, maintenance personnel can check and record actual vibration, pressure-volume, pressure-time, and ignition oscilloscope patterns for each type of reciprocating equipment. Standard oscilloscope patterns which show the normal engine patterns for each engine are compared with those measured by the analyzer in order to assess the engine performance.

In the hands of a skilled operator, the analyzer can improve operating efficiences by early detection of malfunctions such as improper timing, valve leakage, faulty rings, faulty bearings, and ignition-related problems. The operator should be capable of comparing the pattern on display to the standard one and diagnose the probable causes for any abnormality. Guidelines should be established to determine from pattern analysis when maintenance should be performed. Complete files containing standard and actual patterns should be maintained for each unit.*

---

*American Gas Association, *Engine Analyzer Signature Directory* (Arlington, 1973), [Catalog no. XF0478].

## Power Cylinder Load Balancer

Modern power cylinder load balance instruments enable an operator to read the mean effective pressure (MEP) directly from a dial pressure gauge for each cylinder of the engine. The MEP can be balanced evenly between cylinders to improve fuel economy and reduce wear.

Instruments generally used for this purpose measure the cylinder pressure as the gas is either compressed or expanded. Instrument selection will depend on the application and personal preference.

Selection of the proper instrument is important, but more important is implementing a regularly scheduled power cylinder balancing program by a competent operator.

## Optical Aids

The optic level and the borescope are two optical aids that are extremely useful in a maintenance program.

### Optic Level

An optic level is an aligning device used to check for misalignment of reciprocating engine frames. An alignment profile should be established with the optic level for each reciprocating engine main frame. When the main frame becomes misaligned beyond allowable limits, the optic level is utilized for realignment. Alignment should be checked annually and more often if the main frame has a history of misalignment.

The optic level also can be used to set sole plates; to establish main bearing elevations; or to reestablish the profile of the main bearing supports when the crankshaft is removed.

### Borescope

A borescope is a visual inspection instrument which uses a high-intensity light source and flexible fiber bundles to transmit images of the equipment internals onto a screen without disassembly of the equipment. The borescope is used to investigate reciprocating engine problems such as scored cylinders, carbon in ports, intake exhaust and/or fuel valve problems, cracked heads, cracked pistons, etc.

On turbine engines the borescope is used to inspect hot sections and rotating sections inside the turbine. Some turbine manufacturers either provide borescope inspection ports, or offer a retrofit design to accommodate borescope port installation. Pictorial records should be made in problem areas that need a more intensified inspection.

## TESTING AND ANALYSIS

Testing and analysis is an important part of every station maintenance program. An analysis of the lube oil samples, water system and exhaust gas samples, equipment trends and failures should be performed as required.

### Lube Oil Analysis

A wide range of engine lubricants is available. Additive packages are available that contain oxidation inhibitors, rust and corrosion inhibitors, anti-wear and antifoam agents, and other chemicals that are used to improve the properties of engine oils. An engine oil should be selected that meets or exceeds the engine manufacturer's specifications. For special applications or needs, lubrication engineers employed by oil vendors can be consulted. When selecting lubrication oil for reciprocating engines, it should be kept in mind that the lube oil used is very critical for engine maintenance. Economizing on lube oil could result in increased maintenance costs. Experience has shown that using the best lubricating oil on the market is not an expensive choice.

To ensure that the quality of engine lubricating oil is maintained while in use, periodic samples of the used oil should be analyzed. Oil analysis is useful in determining oil contamination, deterioration of oil viscosity, and the rate of engine wear. Oil analysis can be performed by company-owned laboratories, oil vendors, or independent laboratories. The frequency of lube oil analysis will depend on the type of analyses performed, the laboratory facilities selected, the use factor, and the type of equipment.

The most severe use of engine lube oil usually occurs on naturally-aspirated, 4-stroke engines. In this type of engine the air/fuel mixture has a significant effect on oxidation and nitration contamination of the lube oil. Contaminants contribute to a rapid increase in oil viscosity. When the contamination reaches an unacceptable level, the engine oil will have to be drained and replaced.

A tabulation of viscosities of oil used at compressor stations and the maximum allowable value of viscosity should be available to compressor station management.

Two-stroke, supercharged or turbocharged engines and 4-stroke, turbocharged engines are less demanding than naturally aspirated 4-stroke engines. Lube oil contaminants usually occur because of coolant leaking into the crankcase of the lubrication system, or contamination from other foreign agents. It is not unusual to experience 75 000 to 100 000 hours of operation between oil changes on this type of engine, and filters can extend the useful life of the lubricating oil even further. Of course, because of oil consumption new oil is added continually to the crankcase in small quantities.

On late model engines that have a high Brake Mean Effective Pressure (BMEP) rating, analysis of used oil should be more frequent than on the lower BMEP 2-stroke engines. Accelerated wear occurs in high BMEP engines because of higher engine speeds and temperatures.

The oil analysis, in addition to the physical properties test (total solids and water content, viscosity, neutralization number, etc.), should include a spectrochemical test to identify metallic elements contained in the oil sample. Elements, such as iron, chromium, and aluminum, provide a positive and early identification of the critical parts in the engine that are wearing—as the traces of these elements are found long before the actual wear becomes apparent. A sudden increase in the amount of one or more of these elements in the sample is a good indication of an incipient problem. Figure 77 shows a typical engine oil analysis report.

Phosphorus and zinc are part of the antiwear package added to oil. Calcium, barium, and magnesium are the scavengers of the system. It is their function to pick up the contaminants and wear particles, and deliver them to the filter while circulating through the system.

Any dirt, dust, sand, or similar abrasives in oil are reported as silicon. However, silicon could be indicative of an antifoam additive in certain types of oil also; therefore, the data on used oil samples should be compared with that of a new oil. Any sudden increase in the silicon level generally is indicative of a problem in the air intake system.

Sodium borate and sodium chromates are used in coolant systems. In the event of a coolant leak these elements will remain in the lubricant even after the evaporation of the water.

## Water Systems Analysis

Water from the station storage water system, hot water circulating systems, steam boilers, local and domestic makeup water systems, potable water systems, and air compressor water systems should be analyzed on a periodic basis to prevent deterioration and damage to system components due to dirt or contamination.

Fluid samples should be taken every 60 days from the engine jacket water system at the discharge end of the circulating pump. Analysis of the samples should be performed by company personnel or a certified independent laboratory. Tests should be performed to determine the pH, the freezing point, the amount of lube oil contamination, and the

**SPECTRA-CHECK®**

CLEVELAND TECHNICAL CENTER, INC.
13600 DEISE AVENUE
CLEVELAND, OHIO 44110  (216) 451-6455

SPECTRO-CHEMICAL ANALYSIS – ELEMENT CONCENTRATIONS SHOWN IN PARTS PER MILLION BY WEIGHT



CLIENT NO. 571-001   ENGINE MAKE Waukesha   BRAND OIL Mobil 100/40   DET. ☒ YES ☐ NO   FUEL TYPE: ☐ DIESEL ☐ GASOLINE ☒ NAT. GAS ☐ OTHER

UNIT NO. #1 Cambridge   ENGINE MODEL L 7042   ☐ TRANS. ☐ HYD. ☐ DIFF. ☐ GEAR BOX ☐ FRONT ☐ REAR ☐ LEFT ☐ RIGHT ☐ OTHER

AIRCRAFT "N" NO. Station   ENGINE SERIAL NO. 192293   TYPE ANTI-FREEZE OR COOLING SYSTEM INHIBITOR

| TEST NO | LAB NO | DATE TAKEN / TESTED | OIL MI(HR) / ENG MI(HR) | IRON FE | CHROMIUM CR | LEAD PB | COPPER CU | TIN SN | ALUMINUM AL | NICKEL NI | ANTIMONY SB | MANGANESE MN | SILICON SI |
|---|---|---|---|---|---|---|---|---|---|---|---|---|---|
| 1 | 8-9 / 4269 | 8-24 / 8-30 | | 17 | 2 | 68 | 39 | 4 | 1 | 0 | 0 | 2 | 9 |
| 2 | 2-0 / 3374A | 2-22 / 2-25 | | 9 | 0 | 5 | 1 | 3 | 1 | 0 | 0 | 1 | 10 |
| 3 | 6-0 / 2324 | 6-6 / 6-12 | 1239 | 15 | 2 | 5 | 2 | 2 | 2 | 1 | 0 | 2 | 10 |
| 4 | 2-1 / 1322 | 2-2 / 2-6 | 1020:55 | 9 | 0 | 4 | 1 | 2 | 2 | 0 | 0 | 1 | 8 |
| 5 | 9-1 / 2987 | 9-14 / 9-15 | 264 / 24/144:30 | 6 | 0 | 7 | 1 | 1 | 4 | 0 | 0 | 1 | 5 |
| 6 | 4-84 / 1646 | 3-28 / 4-9 | | 7 | 1 | 4 | 1 | 0 | 2 | 0 | 0 | 0 | 4 |

| TEST NO | LAB NO | BORON B | SODIUM NA | MAGNESIUM MG | CALCIUM CA | BARIUM BA | PHOSPHORUS P | ZINC ZN | MOLYBDENUM MO | FLASH (FUEL DILUTION) | VISC @ 100 F / 210 F | % WATER / INSOL | ANTI-FREEZE POS / NEG | PH | TOTAL ACID NO / BASE NO |
|---|---|---|---|---|---|---|---|---|---|---|---|---|---|---|---|
| 1 | 8-9 / 4269 | 1 | 61 | 3 | 868 | 3885 | 78 | 42 | 1 | 420 / <1 | 949.7 / — | 0 / 0.2 | NEG | 3.7 | 6.0 / 8.2 |
| 2 | 2-0 / 3374A | 32 | 82 | 7 | 900 | 4070 | 89 | 63 | 2 | 445 / <1 | 741.1 / — | 0 / Trace | POS | 7.0 | 1.91 |
| 3 | 6-0 / 2324 | 30 | 86 | 8 | 1094 | 3962 | 94 | 88 | 3 | 440 / <1 | 675.7 / — | 0 / Trace | POS | 6.4 | 2.60 |
| 4 | 2-1 / 1322 | 5 | 84 | 6 | 1032 | 3566 | 41 | 75 | 1 | 430 / <1 | 235.6 / — | 0 / Trace | NEG | | |
| 5 | 9-1 / 2987 | 1 | 38 | 3 | 593 | 5178 | 30 | 9 | 1 | 440 / <1 | 632.3 / 73.47 | 0 / Trace | NEG | 7.1 | 0.71 |
| 6 | 4-84 / 1646 | 1 | 0 | 0 | 107 | 545 | 277 | 288 | 0 | 460 / <1 | 638.2 / 72.36 | 0 / Trace | NEG | 7.5 | 1.39 |

TYPE SERVICE I (II) III · ☒ OTHER   ☐ SEE ATTACHED ADVISORY   PHYSICAL TESTS: ☐ NO CORRECTIVE ACTION INDICATED BY TEST PERFORMED   ☐ TEST RESULTS INDICATE OIL CONDITION IS SATISFACTORY   SPECTROCHEMICAL ANALYSIS: ☒ NO CORRECTIVE ACTION INDICATED BY TESTS PERFORMED   ☒ TEST RESULTS INDICATE WEAR METAL LEVELS ARE SATISFACTORY

1  High total acid number and low pH – indicates oxidized oil.  Suggest change lube oil and lube oil filter. (Per Phone)

TYPE SERVICE I (II) III · ☒ OTHER   ☐ SEE ATTACHED ADVISORY   PHYSICAL TESTS: ☐ NO CORRECTIVE ACTION INDICATED BY TEST PERFORMED   ☐ TEST RESULTS INDICATE OIL CONDITION IS SATISFACTORY   SPECTROCHEMICAL ANALYSIS: ☐ NO CORRECTIVE ACTION INDICATED BY TESTS PERFORMED   ☐ TEST RESULTS INDICATE WEAR METAL LEVELS ARE SATISFACTORY

2  Boron and Sodium levels are increasing.  Suggest check engine for internal water leaks. (Per Phone)

TYPE SERVICE I (II) III · ☒ OTHER   ☐ SEE ATTACHED ADVISORY   PHYSICAL TESTS: ☐ NO CORRECTIVE ACTION INDICATED BY TEST PERFORMED   ☐ TEST RESULTS INDICATE OIL CONDITION IS SATISFACTORY   SPECTROCHEMICAL ANALYSIS: ☐ NO CORRECTIVE ACTION INDICATED BY TESTS PERFORMED   ☐ TEST RESULTS INDICATE WEAR METAL LEVELS ARE SATISFACTORY

3  Boron and Sodium levels are remaining high.  Suggest check engine for internal water leaks. (Per Phone)

TYPE SERVICE I (II) III · ☐ OTHER   ☐ SEE ATTACHED ADVISORY   PHYSICAL TESTS: ☒ NO CORRECTIVE ACTION INDICATED BY TEST PERFORMED   ☒ TEST RESULTS INDICATE OIL CONDITION IS SATISFACTORY   SPECTROCHEMICAL ANALYSIS: ☒ NO CORRECTIVE ACTION INDICATED BY TESTS PERFORMED   ☒ TEST RESULTS INDICATE WEAR METAL LEVELS ARE SATISFACTORY

TYPE SERVICE I (II) III · ☒ OTHER   ☐ SEE ATTACHED ADVISORY   PHYSICAL TESTS: ☒ NO CORRECTIVE ACTION INDICATED BY TEST PERFORMED   ☒ TEST RESULTS INDICATE OIL CONDITION IS SATISFACTORY   SPECTROCHEMICAL ANALYSIS: ☒ NO CORRECTIVE ACTION INDICATED BY TESTS PERFORMED   ☒ TEST RESULTS INDICATE WEAR METAL LEVELS ARE SATISFACTORY

TYPE SERVICE I (II) III · ☒ OTHER   ☐ SEE ATTACHED ADVISORY   PHYSICAL TESTS: ☒ NO CORRECTIVE ACTION INDICATED BY TEST PERFORMED   ☒ TEST RESULTS INDICATE OIL CONDITION IS SATISFACTORY   SPECTROCHEMICAL ANALYSIS: ☒ NO CORRECTIVE ACTION INDICATED BY TESTS PERFORMED   ☒ TEST RESULTS INDICATE WEAR METAL LEVELS ARE SATISFACTORY

*Figure 77.* Typical lube oil analysis report

concentration of suspended and dissolved solids in each sample. Figure 78 illustrates a typical engine coolant analysis report.

## Exhaust Gas Analysis

The most efficient engine operation is accomplished when a minimum of free air and carbon monoxide (CO) and a maximum of carbon dioxide ($CO_2$) are present in the exhaust gas stream. Analysis of the exhaust gases permits the setting of a proper air/fuel mixture. Exhaust gas analyses are more difficult to perform on 2-stroke than on 4-stroke, naturally-aspirated engines because of the scavenging air sequence in the 2-stroke engines.

## Trend Analysis

Frequent analysis, testing, and monitoring of prime mover and auxiliary equipment will show trends in various operating parameters. When these trends are compared to the normal operating range of the equipment, future failures can be predicted, and excessive wear avoided. Frequently trended parameters include the engine vibration level, lube oil viscosity and metal content, jacket water and exhaust manifold temperatures, and suction and discharge gas pressures. This informa-

### Coolant Analysis Report

Report Number _______________________

|  |  | Found | Recommended |
|---|---|---|---|
| **Sample Appearance:** | Color | yellow | na |
|  | Oil | none | none |
|  | Appearance | turbid | clear |
| **Corrosion Products:** | Total Iron (ppm as Fe) | 4.9 | 1.0 ppm or less |
|  | Total Copper (ppm as Cu) | .8 | 0.5 ppm or less |
| **Corrosion Inhibitors:** | Orthophosphate (ppm as $PO_4$) | 1900 | 4800 - 7200 |
|  | Nitrite (ppm as $NaNO_2$) | ND |  |
|  | Reserve Alkalinity (ml. of O.1N HCl) | 10.1 | 6.0 - 8.0 |
|  | Boron (ppm as B) | 960 | 80 - 120 |
| **Corrosive Chemicals:** | Chloride (ppm as NaCl) | 110 | 200 ppm or less |
|  | Sulfate (ppm as $SO_4$) | 160 | 300 ppm or less |
| **Scale Forming Chemical:** | Calcium Hardness (ppm as $CaCO_3$) | 6.2 | 0 ppm |
| **Coolant Properties:** | pH | 8.5 | 8.0 - 10.0 |
|  | Ethylene Glycol (%) | 44% |  |
|  | Freeze Point (°F) | -21 |  |

*Figure 78.* **Typical engine coolant analysis report**

tion can be used to prevent equipment failures or loss of performance by replacing the lube oil, oil filters, air filters, and other engine components that eventually could cause the failure.

### Failure Analysis

Premature failure of engine and gas compressor components disrupts any maintenance program. When any major component of a system fails prematurely or unexpectedly, an analysis should be made of the failure to make the following assessment:
- What caused the failure?
- What action should be taken to prevent the occurrence from repeating in similar units?
- What immediate action should be taken to improve the security of those units operating with questionable components?

Failures of components in the high-pressure gas system that could jeopardize station security and cause injuries to personnel warrant immediate attention. Aggravating failures that are not hazardous are less critical.

**NOTE:** *The results of any failure analysis should be shared with all company personnel involved.*

Any significant or repetitive failure of engine components that is caused by a manufacturer design deficiency should be investigated with the intent of persuading the manufacturer to improve the design in order to correct any deficiency and improve the product reliability of future equipment.

### INSPECTION

Several types of inspections can be performed that will enhance a good maintenance program. These include nondestructive testing of critical studs on the engine and measurement of engine component wear with precision instruments.

### Nondestructive Testing

An ultrasonic instrument is vital to a maintenance program. It provides the means for investigating the condition of stud bolts and other heavily loaded fasteners in critical operating areas of both reciprocating and turbine engines. High pressure valve cap stud bolts and other critical fasteners such as head bolts, suction and discharge flange bolts, unloader bottle studs, unit power train bolts, and flywheel bolts, should be tested ultrasonically for flaws either on an annual basis or after approximately 8 000 hours of operation.

On some reciprocating engines the power piston rod studs are critical and are subject to fatigue failure. Operating experience will identify the fasteners that are susceptible to fatigue failure. A program should be established to investigate these fasteners ultrasonically on a periodic basis to prevent catastrophic failure.


**Measurement**

Precision measurements should be made to determine wear trends in reciprocating and turbine engines and compressors. On a routine basis main bearings and rod bearings should be inspected to ensure that proper clearance between the bearings and the crankshaft is maintained. Crankshaft web deflection should be measured with a micrometer on a routine basis or after 8 000 hours of operation to see that it does not exceed the level set by the engine manufacturer. The data should be logged and plotted for each engine. Figure 79 shows a typical crankshaft web deflection log.

Any time a power piston or compressor piston is removed, the piston and the cylinder should be measured with micrometers to determine the wear of these components. Crosshead slides must be checked routinely to ensure that proper clearances are maintained. On accessory gear drive systems, backlash between gears should be measured and recorded annually. Pinion gears or ring gears should be adjusted or replaced as needed to maintain an acceptable backlash.

On turbine engines the manufacturer requires a close tolerance fit for turbine rotors, spaces, guide vanes, turbine cases, bearing clearances, etc. When reassembling, all measurements and clearances must be accurate to allow for thermal growth. All measurements should be recorded in the repair report.


## OPERATIONAL MONITORING

The visual and audible aspects of operational monitoring are inseparable. The plant operators and maintenance personnel must be alert when making routine rounds and look for specific problems. This type of monitoring should be a routine inspection where operators or maintenance personnel look and listen for anything unusual. They should be looking for oil leaks and for unusual movement in pipe supports, compressor foundations, or engine blocks; they should be listening for

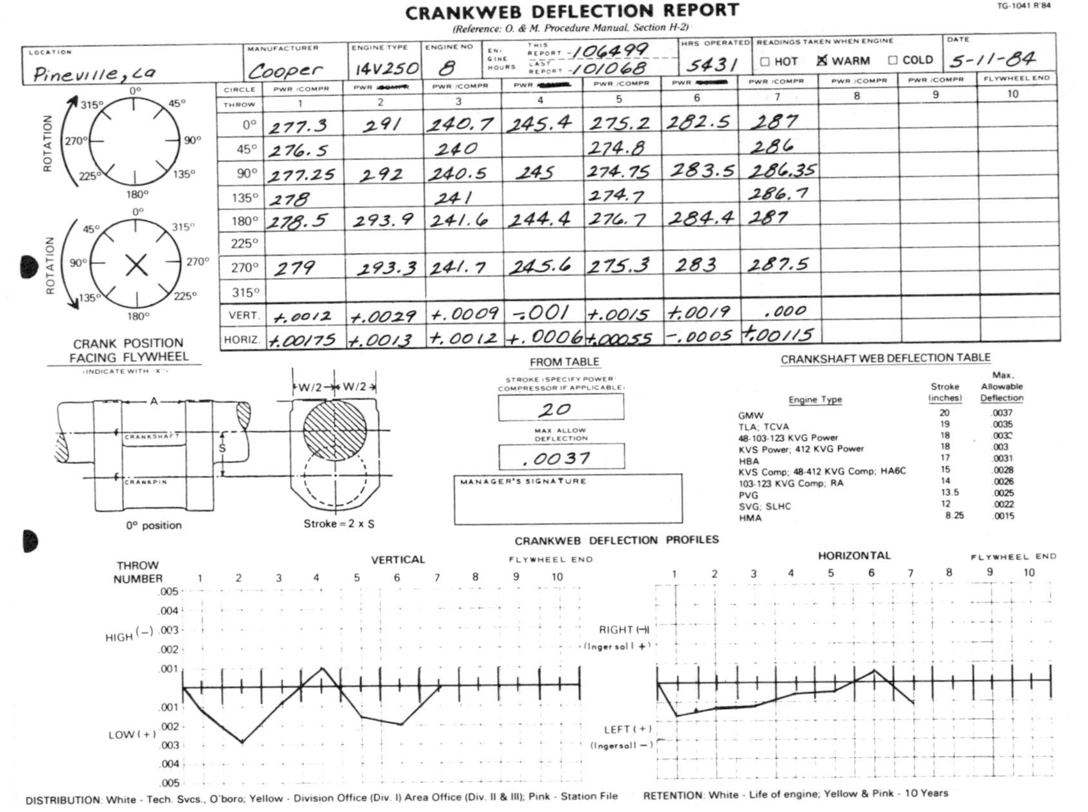

**CRANKWEB DEFLECTION REPORT**

*(Reference: O. & M. Procedure Manual, Section H-2)*

TG-1041 R'84

LOCATION: Pineville, La   MANUFACTURER: Cooper   ENGINE TYPE: 14V250   ENGINE NO: 8   ENGINE HOURS — THIS REPORT: -106499, LAST REPORT: -101068   HRS OPERATED: 5431   READINGS TAKEN WHEN ENGINE: ☐ HOT   ☒ WARM   ☐ COLD   DATE: 5-11-84

| CIRCLE / THROW | 1 PWR/COMPR | 2 PWR/COMPR | 3 PWR/COMPR | 4 PWR/COMPR | 5 PWR/COMPR | 6 PWR/COMPR | 7 PWR/COMPR | 8 PWR/COMPR | 9 PWR/COMPR | 10 FLYWHEEL END |
|---|---|---|---|---|---|---|---|---|---|---|
| 0° | 277.3 | 291 | 240.7 | 245.4 | 275.2 | 282.5 | 287 | | | |
| 45° | 276.5 | | 240 | | 274.8 | | 286 | | | |
| 90° | 277.25 | 292 | 240.5 | 245 | 274.75 | 283.5 | 286.35 | | | |
| 135° | 278 | | 241 | | 274.7 | | 286.7 | | | |
| 180° | 278.5 | 293.9 | 241.6 | 244.4 | 276.7 | 284.4 | 287 | | | |
| 225° | | | | | | | | | | |
| 270° | 279 | 293.3 | 241.7 | 245.6 | 275.3 | 283 | 287.5 | | | |
| 315° | | | | | | | | | | |
| VERT. | +.0012 | +.0029 | +.0009 | -.001 | +.0015 | +.0019 | .000 | | | |
| HORIZ. | +.00175 | +.0013 | +.0012 | +.0006 | +.00055 | -.0005 | +.00115 | | | |

CRANK POSITION FACING FLYWHEEL (INDICATE WITH "X")

0° position    Stroke = 2 x S

FROM TABLE

STROKE (SPECIFY POWER/COMPRESSOR IF APPLICABLE): 20

MAX ALLOW DEFLECTION: .0037

MANAGER'S SIGNATURE

CRANKSHAFT WEB DEFLECTION TABLE

| Engine Type | Stroke (inches) | Max. Allowable Deflection |
|---|---|---|
| GMW | 20 | .0037 |
| TLA; TCVA | 19 | .0035 |
| 48-103-123 KVG Power | 18 | .003C |
| KVS Power; 412 KVG Power | 18 | .003 |
| HBA | 17 | .0031 |
| KVS Comp; 48-412 KVG Comp; HA6C | 15 | .0028 |
| 103-123 KVG Comp; RA | 14 | .0026 |
| PVG | 13.5 | .0025 |
| SVG; SLHC | 12 | .0022 |
| HMA | 8.25 | .0015 |

CRANKWEB DEFLECTION PROFILES



DISTRIBUTION: White - Tech. Svcs., O'boro; Yellow - Division Office (Div. I) Area Office (Div. II & III); Pink - Station File    RETENTION: White - Life of engine; Yellow & Pink - 10 Years

*Figure 79.* **Crankweb deflection report**

unusual noises in the gas compressors, in the engine crankcase area, and in the topworks of the engine.

At attended stations a routine trip through the engine room should be planned so that at least twice daily someone goes over the top of the engine; walks by the compressors; walks through the basement and the auxiliary building; and passes by every piece of operating equipment at the station. Any unusual observation, either visually or audibly, should be reported to the station supervisor.

### Weekly Inspections

Once each week the fire pump and the standby emergency generator should be run for a few minutes to verify that both systems are functioning properly. Also, the voltage on all emergency batteries should be checked and recorded.

Procedures should be developed and implemented for a weekly inspection of the major components of the emergency shutdown system.

### Records and Reports

Maintenance records and reports should be maintained on reciprocating and turbine units. They include such information as repair procedures, wear trends, parts installed, and other pertinent overhaul data. The Records and Reports paragraph in Chapter 3 contains a discussion of the types of records and reports maintained at a station.

# SPARE PARTS

In order to plan for required maintenance and to minimize the time spent making emergency repairs, it is necessary to maintain an adequate supply of spare parts.

Use factors, staging areas, and computer-monitored inventories can be utilized to minimize the spare parts inventory and the associated costs.

### USE FACTOR FOR COMPRESSOR ENGINES

When the use factor of available capacity in a given system is below 60 to 70 percent and the idle engines are interchangeable with those in service, the parts for emergency repair frequently can be secured by cannibalizing an idle engine. The replacement parts inventory should be maintained at a level that is compatible with the use factor.

## STAGING AREAS FOR SPARE PARTS

If similar compressor engines are in service at several different stations and if they have interchangeable parts, staging areas may be set up between stations so that a spare part may be drawn from the staging area inventory when needed.

## COMPUTER-MONITORED INVENTORIES

Computer analysis programs are available that determine the number of parts required in stock, and are based on the following:
- Total number of units (engines) using the part
- Number of failures for the part
- Lead time for procurement of the part
- Number of months from beginning history to present
- Availability factor
- Reliability factor.

This method of keeping an inventory can reduce storage costs effectively while maintaining an adequate supply of spare parts.

*Chapter 5*

# SAFETY AND SECURITY

Safety and security practices at natural gas compressor stations are established and instituted to protect the facility, to ensure the safety of employees and the public, and to prevent accidents that could damage surrounding property. Protection of a facility is necessary to preserve the investment in equipment and integral parts of gas gathering, transmission, and storage systems. Safety and security are improved by installing an emergency shutdown (ESD) system, equipment protection devices, environmental monitoring equipment, fire detection systems, security systems, and by establishing an accident prevention program.

## EMERGENCY SHUTDOWN SYSTEMS

Emergency shutdown systems are installed at compressor stations to protect operating personnel and station equipment during an emergency such as a fire, ruptured pipeline, etc.. The emergency shutdown system shuts down the engine/compressor units, closes the fire gate valves, and vents the station piping. In addition, the system may be required to shut down electrical power and other auxiliary equipment such as pumps, air compressors, boilers, and electrical generators to reduce the possiblity of gas ignition.

Emergency shutdown systems may be activated automatically by sensors that monitor high gas pressure levels, fire, pressure drop, scrubber high liquid levels, explosive mixtures, etc.; manually by compressor station personnel, or remotely by personnel from another location upon receipt of an appropriate alarm.

## EMERGENCY SHUTDOWN SYSTEM COMPONENTS

The emergency shutdown system includes all valves, valve operators, transducers, pilot/power gas piping, electrical circuits, and any other component used to shut in and vent the yard piping. Specifications for the required components and materials usually are identified in various codes that regulate compressor station operation, maintenance, and design. Climate, location, environment, and type of service must be considered when selecting components and materials for the system. It may be necessary to specify anti-corrosive materials or spark-proof equipment in a specific environment or application.

## DESIGN CONSIDERATIONS FOR EMERGENCY SHUTDOWN SYSTEMS

The design of the emergency shutdown (ESD) system will depend upon the size and complexity of the station, the type of activation system (manual, automatic, remote, or a combination of these), and the extent of shutdown required to protect the station personnel and equipment adequately.

In general, the sizing of system components will depend on the time required for the complete sequence of events to occur. Vent piping is usually large enough to allow blowdown in 5 minutes or less. Pilot gas piping and valves should be large enough to allow venting of the pilot system within 1 minute after the start of the shutdown process. Where gear head valves are used, the motor speed and the number of turns required to close or open the valves should be such that there will be sufficient time for normal valve operation.

Ideally the ESD system power source should be independent of the station electrical and/or power gas systems. Both of these systems are subject to interruptions that might interfere with the emergency shutdown sequence.

## TYPES OF EMERGENCY SHUTDOWN SYSTEMS

Although there are many types of ESD systems, they generally fall into two categories: power/pilot gas and electrically operated systems. In some cases air is used rather than gas for the pilot gas; nevertheless, the method of operation is the same.

### Electrically Operated Systems

Electric systems use an ac or dc power source to shut down the engine, to close fire gate valves, and to open station vent valves. The valve operators are driven by electric motors or by some other electrical/power gas arrangement.

Electrical systems are considered less desirable in the gas industry because electrical power must remain in service until all functions of the shutdown have been completed. Uninterruptible power supplies (UPS) provide a means of protection against the loss of power, but they are not a substitute for circuit loss due to insulation failure in a fire.


## Power/Pilot Gas Systems

Gas systems utilize pressurized gas to effect an emergency shutdown. Two separate piping systems are run to each ESD valve. The first system contains pilot gas which usually is maintained at a pressure of 100 lbf/in$^2$ (690 kPa); the second system contains power gas which is maintained at a higher pressure than the pilot gas. The pilot gas pressure, through a system of poppet control valves, is utilized to prevent the power gas from stroking the ESD valves. The ESD system is activated by venting the pilot gas which allows the power gas to operate valves through a direct-acting gas/hydraulic system. Power gas pressure acts against hydraulic fluid which in turn drives a hydraulic motor or piston that operates the valves. This type of system is adapted to valves that have a 90-degree travel range from fully open to fully closed.

Other gas systems use a similar pilot/power gas arrangement, except that power gas is used to drive a gas motor which drives the valve operator. Gas motors are used in systems where several valve turns are required to reach either a fully open or closed position.

The pilot gas pressure also holds the engine fuel gas valve open. During an emergency shutdown, venting of the pilot gas allows the fuel valve to close – shutting down the engines and venting the fuel lines.

Adjusting the pilot gas flow is important. Pilot gas systems are not always free of leaks, and some small amount of gas must be added continually. If compensation is not made, the pilot pressure may decrease enough to generate a false emergency shutdown. The gas flow must be adjusted, within limits, to compensate for small amounts of normal leakage, but not enough to prevent venting the pilot gas system during an emergency shutdown. In most systems trial-and-error adjustments are required to determine how much flow must be provided to the pilot gas system to maintain the correct pressure. Some systems are fitted with adjustable orifice valves which have been sized to maintain the proper flow.

## TESTING, INSPECTION, AND MAINTENANCE OF ESD SYSTEMS

Testing, inspection, and maintenance of an emergency shutdown system should be frequent and thorough enough to ensure that the system will operate satisfactorily under any condition. State and federal codes may require testing and inspection of emergency shutdown systems once each calendar year. In many cases, however, code requirements for testing emergency system operations are a minimum under normal conditions and may not be adequate to ensure a dependable operation under others. Such things as corrosion, climate, seasonal changes, etc., may change operating conditions enough to cause erratic or undependable operation. Therefore, the system may need to be tested, inspected, and adjusted more frequently, and particularly during or immediately following severe climatic conditions. The system should be tested often enough to instill confidence that it will work properly when it is needed.

## RECORDS AND REPORTS FOR EMERGENCY SHUTDOWN SYSTEMS

Records of ESD tests are required to comply with state and federal code requirements. The documentation of system repairs also is helpful to indicate that corrections were made when system failures occurred. Additional reports may be required by regulatory bodies to ensure that the emergency systems are being tested and maintained properly, and are ready for operation at all times.

# EQUIPMENT PROTECTION SYSTEMS

Equipment protection systems should be designed and built to protect all engines, compressors, and associated equipment as well as personnel within the station.

Basic requirements covering equipment protection systems are dictated by the Department of Transportation (DOT), Materials Transportation Bureau (MTB), Office of Pipeline Safety Regulations (OPSR), Title 49, Part 192 (see Appendix A). These minimum requirements are adequate to confine damage resulting from a failure and often can prevent major catastrophes. Because of the escalating cost of equipment repair and replacement, increasing protection to a more sophisticated level is easily justified.

## ENGINE/COMPRESSOR PROTECTION SYSTEMS

Engine/compressor units should have an alarm annunciator to alert to operator and/or a complete shutdown system responding to the following conditions:

- Low lube oil pressure
- Low lube oil level
- High lube oil temperature
- High differential pressure across the lube oil strainer or filter
- Turbocharger overspeed
- High coolant temperature
- Low coolant level
- Excessive wear or failure of bearings
- High bearing temperature
- Engine overspeed
- High winding temperature (electric motors)
- High crankcase pressure
- Voltage, current, and phase loss (electric motors)
- Restricted air intake
- Low control air pressure
- Inadequate lubricant distribution
- Excessive engine torque
- Fire detection
- Combustible gas detection
- High compressor cylinder temperature
- High compressor discharge pressure
- Excessive vibration
- Low compressor suction pressure

Upon recognition of any one of these conditions, the protection system should indicate an alarm or automatically shut down the unit and isolate it. The degree of protection and number of safety shutdown devices used on a particular unit generally are related directly to the value of the equipment, the nature of operating mode such as whether the station is attended or unattended, and whether the operating history indicates that a particular problem is prone to occur with the specific type of equipment or operation.

### Shutdown Philosophy for Engine/Compressor Units

There are two types of systems available for the shutdown of an engine/compressor unit. In the first system the unit shuts down whenever an equipment protection sensor indicates that a particular key parameter is outside the normal range. This type of system may produce frequent nuisance-type shutdowns due to momentary excursions beyond the normal operating range.

In the second system an alarm set point and a shutdown set point are assigned to each sensor. When the alarm set point is reached, the sensor generates an alarm indicating that the unit is operating beyond its normal, but still within acceptable, operating limits. This provides the operator with the opportunity to inspect and correct the problem before the unit shuts down. If the problem cannot be corrected and the unit continues to operate until it reaches the shutdown set point, it will shut down automatically. Alarm/trip systems generally are preferred to single-trip shutdown systems.

## Optical Devices and Annunciators

Optical devices and annunciators are used in an engine/compressor protection system to aid the operating personnel in detecting either an abnormal operating condition that is developing or a condition that already has occurred and may have caused a shutdown. Optical devices most often used are level gauges and other visual indicators, either pneumatically or electrically operated, that are associated with an individual shutdown sensor or system. Normally located in a main engine control panel, the shutdown indicators provide a convenient means of determining the cause of a shutdown, thereby reducing investigative time to a minimum.

## Equipment Protection Sensors

Although each sensor is designed to respond to the excursions of the variable being monitored, the failure of one sensor is not enough to void the protection of a unit in a well-designed system. For example, if coolant were leaking into an engine crankcase and the coolant level became dangerously low, the high coolant temperature shutdown probably would shut the engine down before serious damage could occur. Should this shutdown fail, the coolant leak probably would contaminate the lube oil and cause the lube oil differential pressure shutdown to activate. If the oil filter differential pressure shutdown failed to function, the low lube oil pressure, high lube oil temperature, high bearing temperature, excessive engine torque, or vibration shutdowns would be activated, depending upon which shutdown mechanism first sensed operating conditions outside the acceptable range. No matter what system of protection is used, it is always best to detect an operating malfunction early to minimize damage.

## OTHER EQUIPMENT PROTECTION SYSTEMS

Other equipment protection systems include overpressure protection devices such as relief valves or monitoring regulators; engine exhaust muffler vents; air intake, backfire pressure, relief valves; and combustible gas atmosphere or flame detectors. These are covered in more detail in Environmental Monitoring Systems.

## TESTING AND MAINTENANCE OF PROTECTION SYSTEMS

Testing and maintenance of protection systems are necessary if reliability is to be maintained. The DOT Code requires each operating organization to have effective operating procedures and a program for maintaining and testing the protection equipment. Operating and shutdown procedures should include all systems associated with the operation of the engine/compressor units.

In addition to shutdown procedures, the DOT Code requires that equipment be isolated for safety during maintenance and alterations. When performing maintenance or alterations to the equipment and protection systems, hazardous conditions may exist which must be controlled. When these conditions develop, the protection equipment should activate the shutdown system unless that section of the protection system is rendered inoperative. These features must be considered when the protection systems are being designed.

### Testing of Protection Systems

Since turbines and centrifugal compressors operate at speeds much greater than reciprocating units, failure and damage usually occur in a much shorter time period; consequently, it is advisable that testing and inspection of turbine protection systems be done more frequently than is done for reciprocating units.

Hydrostatic testing of station piping is necessary to verify the integrity of the system and reduce the possibility of failures initiated by leaks or a rupture in the piping system.

Automatic valve or regulator failures could result in overpressure of the piping system in compressor stations, particularly immediately downstream of the compressor unit. Therefore, it is vital that pressure relief devices and safety shutdown systems relating to overpressure be tested and inspected regularly and that they be maintained in first-class condition.

## Inspection of Engine/Compressor Units

Even though they are tested to a standard such as the American Society of Mechanical Engineers Power Test Code (ASME Test) at the factory, engine/compressor units should have field tests performed to ensure that the specified performance parameters are attained. Such a field test also will determine the performance of the unit when integrated into the compressor station piping system.

When an engine/compressor unit is shipped to the compressor station site, in many cases the equipment is not received in the same condition in which it was shipped. Units large enough to require shipment by rail require close monitoring of the shipping route as well as careful handling during shipment. The manufacturer usually takes precautions to secure the unit aboard a rail car, but is not responsible for the actual transportation by rail. Accelerometers often are attached at critical locations to monitor the loaded railroad car in the event damage is discovered after shipment and prior to start-up.

All the necessary caution notices will not prevent careless or inadvertent damage to equipment in transit. This is especially true of equipment containing ball or roller bearings such as aircraft-derivative turbine engines. Damaged bearings that cannot be detected, or bearings that appear to be undamaged can fail prematurely during the first hours of operation, and usually result in costly repairs. Protection from the elements and adverse weather conditions during shipment of an engine/compressor unit is absolutely necessary if this type of failure is to be prevented. Moisture and dust are perhaps the most damaging contaminants encountered. Inevitably these contaminants are present during shipment and on location during the construction period prior to the initial operation.

## Records and Reports of Protection Systems

Records and reports of all tests performed on equipment protection systems should be kept on file at the station to verify that the systems are maintained properly and are in a reliable working condition.

# ENVIRONMENTAL MONITORING SYSTEMS

Environmental monitoring systems are used in compressor stations primarily to detect the presence of combustible gas, or to detect flames and activate equipment shutdown and/or fire extinguishing systems.

## COMBUSTIBLE GAS DETECTION EQUIPMENT

Gas detection equipment falls into one of two categories. The first type siphons atmospheric samples from several locations, and analyzes the explosive potential of each sample sequentially. This system operates slowly because of the time required for the sample to travel to the analyzer instrument location. Also, the accuracy may be impaired because of the dilution effect on the latest sample by mixing with the previous one in the sample delivery system.

The second category employs one or more diffusion sensor assemblies at each area being monitored. The sensor produces an electrical signal proportional to the concentration of the combustible gas. The signal is amplified by a measurement module which may incorporate an indicating meter and one or more signal relays generally tripped by the malfunction of equipment components – first warning (low level concentration) alarm followed by a danger warning alarm. Since ventilation is a code-recognized means of coping with explosive atmospheres in a given area, a first warning relay sometimes is used to start supplementary ventilation systems.

The diffusion sensors may use the principle of catalytic combustion of flammable gas or solid state gas detection using a metal oxide silica element. In catalytic combustion a heated catalyst is exposed to the combustible gas or vapor which is oxidized on the surface of the catalyst. Oxidization causes heat, and the temperature rise of the sensing element is related directly to the concentration of the combustible.

The solid-state detector does not operate on the combustion principle, and the presence of oxygen is not required for gas detection. Detection of gases is accomplished by imbedding a heater and collector in a metal oxide silica type of material. Its temperature is fairly low, and high resistance exists between the sensor elements. Upon sensing gas this resistance changes producing a large signal to drive the measurement module. The sensor is relatively immune to poisoning by carbon monoxide or the halogenated compounds; therefore, both its shelf life as a spare and the service life in the field are claimed to be considerably superior to catalytics.

The acronym LEL (or LFL) has become associated with gas detection equipment, and simply implies that the atmosphere contains a certain percentage of the lower explosive or flammable limit (LEL or LFL) of the specific gas/air mixture. For example, a 30 percent LEL of methane/air mixture, which has an explosive range of 5-15 percent, would contain 1.5 percent (30 percent of 5 percent) of methane in the atmosphere.

## Location of Gas Detection Equipment

Gas detection equipment should be used to monitor any location with a potential exposure to natural gas leakage. Typical locations include those areas near compressor cylinders, fuel manifolds, gas heaters or boilers, and gas metering or regulating facilities.

## Alarm/Shutdown Philosophy for Gas Detection Equipment

The alarm and shutdown philosophy for each gas detection system varies according to the operation of the facility. Manned stations can use the warning or the alarm signal to alert personnel and in some cases to start air-circulating fans in the building automatically. A higher explosive mixture condition, which is sensed by a high LEL (LFL) set point, can initiate the shutdown of the equipment near the gas leak.

Unmanned facilities employ combustible gas detection circuitry which starts the ventilation fans upon the first alarm, and if the gas content continues to increase, trips the station emergency shutdown system when the high LEL set point is reached.

## FIRE DETECTION AND EXTINGUISHING SYSTEMS

Fire detectors and fire extinguishing equipment are installed at compressor stations to prevent injury to operating personnel and damage to equipment.

## Types of Fire Detection Equipment

Optical devices, such as ultraviolet or infrared systems, heat sensors, or smoke detectors, can be utilized to detect fires within the compressor or auxiliary buildings. These devices can be wired to activate alarms automatically, initiate an emergency shutdown, and/or activate fire extinguishing systems.

## Fire Extinguishing Equipment

Foam, halon, dry chemicals, and carbon dioxide extinguishing systems can be utilized to prevent extensive damage to compressor station facilities in the event of a fire.

## Fire Detection and Extinguishing Philosophy

Extinguishing systems can be either automatically operated by a controller or manually operated by station personnel.

The amount of protection usually depends upon the size of the facility, the type of operation (manned or unmanned), and the importance of the facility to the system. It should be kept in mind, however, that any system is subject to malfunction and that a fire extinguishing system that is released automatically by a flash of lightning or some other erroneous signal can become a serious nuisance because of the clean-up involved.

## SECURITY SYSTEMS

Security systems available for compressor stations may include automatic gates, floodlights, intrusion alarms, closed circuit television cameras, and guard services. The objective of security systems is to prevent intrusion by unauthorized persons, theft, and vandalism.

## ACCIDENT PREVENTION

Accidents that might result in minor or serious injuries to station personnel can be prevented by establishing an accident prevention program. The program should include the following:

- Written procedures for assembling and disassembling station equipment
- Written instructions on the proper use of tools and equipment
- Inspection of tools for defects and wear before and after use
- Posting "No Smoking" signs in gas-susceptible areas of the station
- Maintaining orderly work areas
- Instructing new employees of the potential hazard of high pressure gas
- Frequent review of emergency operating procedures.

Confined areas should be checked periodically to be certain that the surrounding air is safe to breathe, and all sources of gas leakage are controlled or eliminated. No matches, smoking, or open flames should be permitted in or near the compressor building, in areas classified as hazardous per the NEC, or where a flammable atmosphere may exist.

Accident prevention programs are only effective when they are applied conscientiously. Therefore, it is the station manager's responsibility to see that all personnel observe all safety rules and regulations. Visitors and contract personnel should be instructed in all safety requirements and be required to follow program directives.

# EMERGENCY PROCEDURES

As discussed under Standard and Emergency Operating Procedures, a written plan of action for handling an emergency situation in a compressor station is required by the DOT regulations. This procedure should identify the course of action necessary to isolate station facilities in order to reduce the area that could be affected by the condition creating the emergency situation.

Personnel should be trained to execute the plan. Periodic meetings should be conducted to ensure that everyone is familiar with the proper technique in handling emergencies. Included in the training sessions should be instructions in fire fighting and methods of administering first aid.

**NOTE:** *Emergency and alternate emergency gathering areas should be designated so that personnel can be accounted for in the event of any emergency. Visitors to the facility should be made aware of the whereabouts of gathering areas.*

*Chapter 6*

# ECONOMIC EVALUATION

Economic alternatives always should be considered in the planning, operation, or maintenance of a compressor station. Regardless of the regulatory body responsible for approving expenditures or rate increases for the operating company, the ultimate objective for each company should be to provide the best possible service at the minimum cost to the consumer. Frequently economic evaluations include not only the compressor station but also other portions of the natural gas system such as pipelines, production, gathering, storage, and supply/market strategies.

There are various ways of making economic comparisons, but all require that certain cost information be available or predictable. In addition, it is important to consider several different alternatives that effectively will achieve the same objective. Once two or more plans for a particular program are identified and a method of estimating the costs associated with each plan is determined, the planning engineer or the operating manager can select whichever method of analytical evaluation meets the organizational needs.

Since the subject of economics is treated in more detail in another book of the GEOP series, this chapter will highlight only a few of the most common aspects of compressor station planning related to economic evaluation. The first discussion explores the difference between capital investment cost and annual cost. The second discusses the option of leasing as compared to purchasing compressor station equipment. Finally, the third points out major factors that should be considered in making an evaluation of comparable equipment.

One method of comparing economic alternatives is the Uniform Annual Equivalent Revenue Requirements (UAERR) method. Simply stated, this method takes into consideration the primary factors that affect all economics-related principles within any major organization.

These include the minimum acceptable return on equity, debt ratio, corporate tax structure, rate and method of depreciation, and interest on borrowed money.

The Revenue Requirements approach is a method of determining the revenues that must be generated from a particular investment in order to cover all of the costs including the required rate of return (return of and on investment). Revenue requirements also may be used to compare alternatives. The alternative that produces the least revenue requirements is usually the most economical. Ordinarily compressor station economics does not involve revenue as such, but rather an expense to the operator which can be considered an annual cost. Treated as a revenue requirement, the annual cost includes the cost of the investment annualized over the life of the facility, plus the annual operating and maintenance expenses. When comparing alternatives, the one which produces the lowest annual cost is the most economical.

Another case in which economics should be considered in compressor station operations is where an existing facility or type of operation could be modified or replaced with a more efficient one. The net savings in the annual cost then are equivalent to the generated revenue, and if the annual savings are greater than the revenue requirements, the project usually is justified. These examples, of course, are somewhat oversimplified in that most compressor station planners are faced with a combination of factors. Generally, whenever a new compressor station is being considered, other criteria—such as location, pipeline sizes, manpower availability for operation, and environmental impact- must be considered. Occasionally, where additional capacity to an existing station is being considered, it may be more economical to replace the original equipment with entirely new, more efficient equipment with a greater capacity; whereas the replacement of the existing equipment alone may not be justifiable.

As with any economic decision, successful compressor station planning depends entirely upon the ability to accurately project costs and performance. Since this is a rapidly changing technology, it would be misleading to list current criteria such as cost per unit of power for a particular type of installation, or fuel rates for various types of prime movers. The planner should seek current data for each project in order to make a complete evaluation of the various available options.

# CAPITAL INVESTMENT VERSUS ANNUAL COST

Two methods of evaluating alternatives in the planning process are the capital investment method and the annual cost method. Occasionally the two yield the same results; however, most often they produce entirely different solutions if enough alternatives are available.

## CAPITAL INVESTMENT METHOD

The capital investment method is simply a method of comparing several alternatives, each of which will satisfy the specific requirements, and selecting the one that requires the least capital investment. Although this method is not popular in industry today, it is useful where minimizing the initial investment is important. Many compressor repair projects are based on the capital investment method combined with good engineering judgment.

In some cases it is necessary to install a facility to satisfy a temporary or intermittent need such as a seasonal peaking operation. In this instance it may be desirable to minimize the investment cost at the expense of the operating costs since they may vary from year to year. Another example where the capital investment method may be preferred involves the rate structure and current financial situation of the company. It may be beneficial to minimize the capital investment because of available sources of revenue or time restrictions imposed on completing the project.

When the capital investment method is used, the planning engineer should explore a variety of alternatives in selecting the equipment and the type of station design. Some of the considerations are:

- Number of Compressor Units
- Type of Prime Movers
- Type of Compressors
- Type and Size of Structures
- Overall Construction Costs
- Installation Schedule

## ANNUAL COST METHOD

The annual cost method is a good method to determine the most economical long-term decision assuming the operating and maintenance costs are reasonably predictable. This method combines the annualized value of the capital investment with the present worth of the annual operating and maintenance costs, and treats the total amount as an annual cost of owning and operating the facility.

In addition to computing the annual cost of the investment, which is the Uniform Annual Equivalent Revenue Requirements factor multiplied by the investment, the cost of operating and maintaining the facility must be estimated for each alternative. The predominant operating cost of a compressor station traditionally has been fuel costs; consequently, emphasis needs to be given to estimating the fuel consumption as well as the fuel cost. If the fuel cost is expected to change at a rate different than the normal rate of inflation, it may be necessary to apply a gradient factor to the fuel costs even though all other operating and maintenance expenses are expected to increase at a constant rate. Another approach is simply to prepare a year-by-year estimate of the total operating and maintenance expenses for the life of the plant and apply a present worth factor to each year.

The purpose of using the annual cost method is to compare the total relative cost of owning and operating each proposal that will meet the requirements. Generally the alternative that has the lowest total annual cost is the most economical in the long term.

## LEASING COMPRESSION EQUIPMENT

Until recently nearly all compressor station equipment was purchased, and became part of the user's equity for financial and taxing purposes. Many vendors now offer compressor units on a lease or lease/purchase arrangement.

Many users have found the lease program financially attractive, particularly in times of economic instability or in a situation where the useful life of the facility is uncertain. For example, gas gathering compressors often are required for only a short period, or when the gas supply is uncertain. In this case the user may elect to lease the equipment with an option to purchase it once the need becomes more definite.

## EVALUATION OF EQUIPMENT

The success or failure of an economic evaluation primarily depends upon how accurate the evaluation is made of the major equipment, particularly the prime mover and compressor units. This evaluation process can become complicated, and often it is experience alone that prompts a specific type of design or equipment. Companies tend to develop a definite pattern in the design of a station, usually as a result of prior experience. Consequently, the planner frequently is influenced toward certain types of equipment; and the alternatives that are considered for evaluation are limited substantially.

The evaluation process normally begins with certain known conditions such as suction pressure, discharge pressure, rate of flow, and geographic location. Though additional criteria is necessary, these basic conditions are enough to begin the evaluation.

Once the compression power requirements are known, it is necessary to consider the various makes and models, and the number of units which would satisfy the operating characteristics of the system. It is important to consider the latest available technology in order to make a thorough comparison of the various types of equipment. A list of alternatives should include the following information:

- Prime Mover (Engine, Turbine, Electric Motor)
- 2-Stroke, 4-Stroke Engines
- Simple Cycle, Regenerated Turbines
- Turbocharged, Naturally-Aspirated Engines
- Reciprocating, Centrifugal Compressors
- Thermal Efficiency
- Fuel Rate (Gas, Electricity)
- Cost of Installation
- Manufacturer's Credibility
- Operating Mode (Automatic, Manual)
- Manpower Requirements
- Life Expectancy
- Ancillary Equipment
- Maintenance Costs
- Parts Availability
- Operator Experience and Acceptance

Each of these items is considered intuitively by an experienced compressor designer, and usually carries a relative rating for each available type of equipment. Coupled with company philosophy, these ratings usually reduce the potential list of alternatives to no more than five or six. Manufacturers' proposals often are solicited so that a detailed analysis can be performed.

Because of rapidly changing technology it would be misleading to make any definite statements relative to an assessment of current equipment. For example, at one time turbine prime movers were less expensive to install than reciprocating prime movers, but the fuel rate for turbines was significantly greater than for equivalent reciprocating engines. This comparison no longer may be valid. Another variable is the construction costs associated with different types of designs. In one case a shop-fabricated unit may be more economical, whereas in another case a field-erected unit with separate ancillaries may be preferable.

Equipment evaluation is perhaps the most difficult and the most critical task involving the economic analysis of compressor station design. There is no simple formula or shortcut if good, sound results are expected. As with any economic study, quality and cost are related directly, and it is important to consider this relationship. Extra effort in estimating the installation costs, operating performance, and operating and maintenance costs will be rewarded many times if the final selection proves to be the correct one.

# EXAMPLES OF ECONOMIC ANALYSIS

The application of economic principles in compressor station planning normally consists of considerable design and engineering effort prior to performing the actual engineering economics study. Accurate cost estimating for the installation, as well as the operating and maintenance expenses, is essential to effective economic evaluation. Once the preliminary engineering for several alternate proposals is completed, the economic evaluation is usually quite simple. The following example illustrates the method of Uniform Annual Equivalent Revenue Requirements in comparing three different plans for a proposed compressor station, each of which would satisfy the requirements of the program.

*Plan 1:* This plan includes one 5 000-hp (3 730-kW) reciprocating, 2-stroke, integral compressor unit designed for an unattended manual mode of operation.

*Plan 2:* This plan includes four 1 350-hp (1 007-kW) reciprocating, 4-stroke, separable compressor units designed for an unattended automatic mode of operation.

*Plan 3:* This plan includes two 3 500-hp (2 611-kW) simple cycle turbines with centrifugal compressors designed for an unattended automatic mode of operation.

The life expectancy of all three plans is 25 years, and the UAERR factor is 0.18. For the first economic comparison (Case 1) the cost of fuel gas is \$3.75/Mcf (\$0.132/m$^3$), and is expected to increase over the life of the plant at the same rate as all other costs.

Table 5 lists the estimated installation and operating costs, and shows the total equivalent annual costs for the three plans. From this data it can be seen that *Plan 2* is the least costly, and would result in an annual savings of \$26 000 as compared to *Plan 3*, and \$50 000 as compared to *Plan 1*.

For the second case it is assumed that the cost of the fuel is expected to increase at a constant rate which is 2 percent greater than the normal rate of inflation. The equivalent annual cost of fuel then would be \$3.75×(1+.02$t$) per Mcf [\$0.132×(1+.02$t$) per m$^3$], where $t$ is the number of years after the installation. Therefore, the present worth of the fuel rate after 1 year would be \$3.83 per Mcf (\$0.135 per m$^3$); after 10 years the rate would be \$4.50 (\$0.159); and after 24 years the rate would be

### TABLE 5
### Proposed Compressor Station Economic Evaluation

| Description | Plan 1 | Plan 2 | Plan 3 |
|---|---|---|---|
| Number of Units | 1 | 4 | 2 |
| Unit Power Rating, hp | 5 000 | 1 350 | 3 500 |
| (kW) | (3 730) | (1 007) | (2 611) |
| Total Power Rating, hp | 5 000 | 5 400 | 7 000 |
| (kW) | (3 730) | (3 021) | (5 222) |
| Total Installed Cost | $11 250 000 | $10 000 000 | $7 700 000 |
| Annual Operating Cost | $150 000 | $125 000 | $75 000 |
| Annual Maintenance Cost | $200 000 | $275 000 | $150 000 |
| Annual Fuel Consumption, Mcf | 280 000 | 340 000 | 504 000 |
| (m³) | (7 929 000) | (9 628 000) | (14 272 000) |
| Annual Fuel Cost | $1 050 000 | $1 275 000 | $1 890 000 |
| Annual Cost of Investment (Installed Cost × UAERR Factor) | $2 025 000 | $1 800 000 | $1 386 000 |
| Total Annual Operating and Maintenance Cost | $1 500 000 | $1 675 000 | $2 115 000 |
| Total Annual Cost (Case 1) | $3 525 000 | $3 475 000 | $3 501 000 |

$5.55 ($0.196). To emphasize, these rates are the present worth values, and should not be confused with the actual cost at a future date.

Since the rate of increase follows a straight line gradient, the average present worth cost for the entire 25-year life may be treated as a uniform annual cost for the entire period. Therefore, the average present worth value is equal to ($5.55 + $3.75)/2, or $4.65 per Mcf [($0.196 + 0.132)/2, or $0.164 per m³]. In this case the total annual fuel costs would be as follows:

*Plan 1:* $1 302 000

*Plan 2:* $1 581 000

*Plan 3:* $2 343 600

The total annual costs would be:

*Plan 1:* $3 677 000

*Plan 2:* $3 781 000

*Plan 3:* $3 954 600

Under these conditions *Plan 1* is the most economical choice, and results in an annual savings of $277 600 as compared to *Plan 3*.

Finally, in Case 3 it is assumed that the cost of gas will increase at a rate that is 1 percent less than the rate of inflation. Using the same method as in Case 2, the average present worth of the cost of fuel would be $3.30 per Mcf ($0.117 per m³). Again, the total annual fuel costs would be:

*Plan 1:* $  924 000
*Plan 2:* $1 122 000
*Plan 3:* $1 663 200

The total annual costs would be:

*Plan 1:* $3 299 000
*Plan 2:* $3 322 000
*Plan 3:* $3 274 200

In this example, *Plan 3* is the best choice with an annual savings of $24 800 as compared to *Plan* 1, and $47 800 as compared to *Plan 2*.

Of course these examples were selected to demonstrate the sensitivity of the total fuel cost in performing economic analyses. It also is apparent that predicting other operating and maintenance costs is important, which in some cases can be predicted more accurately than the fuel cost. For example, equipment maintenance expenses often are low for the first several years of operation, but then increase at a rapid rate as the equipment becomes older. Another consideration is that maintenance costs may be periodic. For instance, turbine units may have a low annual maintenance cost for 4 years, then the cost increases sharply in the fifth year because of a complete overhaul; and the cycle continues throughout the life of the unit.

## SUMMARY OF ECONOMIC EVALUATION

Economic evaluation in the engineering, design, operation, and maintenance of compressor station facilities may range from a decision to replace an electric motor on a gas aftercooler to the selection of pipe size and compressor power rating for a major gas transmission project. In the case of the motor replacement, the cost of installing a new motor is compared with the expected savings in operating and maintenance costs. The solution usually is straightforward and accurate. Decisions of this type are made with little difficulty and risk; however, the secret to making the decision is in recognizing that there are choices available to the operator rather than simply waiting for an old, inefficient motor to wear out. Too often these choices go unnoticed, and an inefficient operation continues indefinitely while potential savings in operating costs are lost.

At the other end of the spectrum, a planning engineer may be evaluating a new gas supply project involving several hundred miles of pipeline and thousands of units of power for a compressor station. In this instance the choices are infinite, and the planner must be capable of selecting those pipeline and compressor sizes that are economically practical and operationally effective. As illustrated in the examples given earlier, the various types of compressor station configurations may render entirely different economic results even though the capacity of the station is constant. When a pipeline is involved in conjunction with a compressor station, the number of alternatives is increased considerably. In this case, because of corresponding changes in pressure conditions, the power rating of the compressor station varies for each increment of size the pipeline is changed on either the suction or discharge side of the station.

While the economic evaluation is simple in the case of the motor replacement, it is extremely complex in the case of the gas supply project. The decision in the first case is simply a yes-or-no decision depending upon the operating cost savings with the new motor. In the second case the economic objective is to determine the combination of pipeline sizes and compressor station configuration that would minimize the cost of service to transport the desired volume rate for the required period of time.

Although estimating the installation costs for various types of compressor stations is important, accurate projections of operating and maintenance expenses, particularly fuel costs, is even more critical in making correct economic decisions. Since the fuel expense of a compressor station represents as much as 75 percent of the total operating cost, not only fuel consumption but projected fuel cost throughout the life of the facility is vital to good planning.

Finally, once the decision is made to install a specific system, the economic evaluation of the system does not end. As implied in the earlier example of the motor replacement, changing technology provides ongoing challenges for planners and designers in the replacement or modernization of existing facilities. Occasionally within a year or two after a new station is put into service, new equipment (such as more efficient compressor valves or waste heat recovery systems) becomes available to improve the operation. It should be understood that even though a particular decision was made with the information available at the time, conditions and technology are changing constantly. Therefore, each opportunity that is present, regardless of the age of the equipment, must be analyzed independently and without consideration of previous decisions. Engineering economics provide the necessary tools to quantify and justify those adjustments and modifications.

# CODES AND STANDARDS

Codes and standards governing the design, construction, and maintenance of compressor station facilities are found in publications issued by various regulatory and governing agencies.

It is not the intent of this Book to provide a comprehensive explanation of each code that is applicable to compressor station design and operation, but instead provide a listing by agency and professional society of all codes that may apply to compressor stations.

Below is a list of codes and standards established by various federal, state, and other agencies. When a question concerning design or operation arises, the engineer can refer to this listing to determine which codes and standards apply to the point in question.

U.S. Government regulations are printed in the Code of Federal Regulations (CFR) published by the U.S. Government Printing Office. Mail orders should be addressed to: Superintendent of Documents, U.S. Government Printing Office, Washington, DC 20402; phone orders may be placed by calling (202) 783-3238.

## U.S. DEPARTMENT OF TRANSPORTATION (DOT), MATERIALS TRANSPORTATION BUREAU (MTB), OFFICE OF PIPELINE SAFETY REGULATION (OPSR), PART 192, TITLE 49 OF THE CODE OF FEDERAL REGULATIONS

### DOT Code, Part 192, Title 49, Subpart C—Pipe Design

- Section 192.111—Design Factor (F) for Steel Pipe
- Section 192.115—Temperature Derating Factor (T) for Design of Steel Pipe

**DOT Code, Part 192, Title 49, Subpart D—Design of Pipeline Components**

- Section 192.163 – Compressor Station – Design and Construction
  - Location of Compressor Building
  - Building Construction
  - Exits
  - Fenced Areas
  - Electric Facilities
- Section 192.165 – Compressor Stations – Liquid Removal
- Section 192.167 – Compressor Stations – Emergency Shutdown
- Section 192.169 – Compressor Stations – Pressure Limiting Devices
- Section 192.171 – Compressor Stations – Additional Safety Equipment
- Section 192.173 – Compressor Stations – Ventilation
- Section 192.201 – Required Capacity of Pressure Relieving and Limiting Stations

**DOT Code, Part 192, Title 49, Subpart J—Test Requirements**

- Section 192.505 – Strength Test Requirements for Steel Pipeline to Operate at a Hoop Stress of 30 Percent or More of SMYS

**DOT Code, Part 192, Title 49, Subpart M—Maintenance**

- Section 192.729 – Compressor Stations: Procedures for Gas Compressor Units
- Section 192.731 – Compressor Stations: Inspection and Testing of Relief Devices
- Section 192.733 – Compressor Stations: Isolation of Equipment for Maintenance or Alterations
- Section 192.735 – Storage of Combustible Materials
- Section 192.751 – Prevention of Accidental Ignition

NOTE: *The above DOT Regulations are also published in the ASME Guide for Gas Transmission and Distribution Piping Systems (see Bibliography).*

<br>

### U.S. DEPARTMENT OF LABOR
### OCCUPATIONAL SAFETY AND HEALTH
### ADMINISTRATION (OSHA)

**See Code of Federal Regulations, 29 CFR Parts 1900–1990.**

# U.S. ENVIRONMENTAL PROTECTION AGENCY (EPA)

## Regulations under the Clean Air Act as amended

- 40 CFR 50 – Air Quality Standards
- 40 CFR 60 – New Source Standards of Performance for Turbines and Reciprocating Engines
- 40 CFR 51 and 52 – Prevention of Significant Deterioration (Best Available Control Technology) in Attainment Areas
- 40 CFR 51 – Visibility Protection – The prevention of any future, and remedying of any existing impairment of visibility in Federal Class I areas.
- 40 CFR 51 and 52 – Requirements for Offsets and Lowest Achievable Emission Rates for New or Modified Sources in Non-attainment Areas

## Regulations under the Clean Water Act of 1977 as amended

- 40 CFR 112 – Spill Prevention, Control and Countermeasure Plan for Oil and Hazardous Substances Storage
- 40 CFR 125 – National Pollution Discharge Elimination System for Waste Water Discharges to Surface Waters

## Regulations under the Resource Conservation and Recovery Act as amended

- 40 CFR 262 – Standards for Generators of Hazardous Wastes
- 40 CFR 264 and 265 – Standards for Hazardous Waste Treatment, Storage, and Disposal Facilities

# LOCAL CODES AND ORDINANCES

Facilities within incorporated areas are regulated by individual city, municipal, county, and state codes administered by various agencies including, but not limited to, the following:

## Department of Buildings and Safety (City)

- General Building Requirements and Permits
- Zoning Requirements
- Grading Requirements and Permits

### Building and Safety Commission

- Electrical Requirements and Permits
- Heating and Air Conditioning Requirements and Permits
- Plumbing Requirements and Permits
- Boiler and Pressure Vessel Inspections and Permits
- Elevator Inspections and Permits

### Department of Public Works

- Environmental Impact Reports
- Driveway and Sidewalk Requirements and Permits
- Sewer Requirements and Permits
- Excavation Permits

### Department of Water and Power

- Electrical Service
- Water Service

### Planning Commission

- Zone Changes

### County Building and Safety (Unincorporated Areas)

- General Building Permits
- Grading Permits
- Electrical Permits
- Plumbing Permits
- Compaction Requirements

### County Engineer

- Geological Reports
- Sewers

### County Regional Planning (Unincorporated Areas)

- Zoning Requirements
- Environmental Impact Studies

## County Air Pollution Control District

- Engine Exhaust Emissions
- Tank and Pond Vapors
- Other Discharges Into the Atmosphere

## County Flood Control District

- Water Discharges

## County Health Department

- Septic Systems
- Water Wells

## Public Utilities

- Compressor Station Design and Construction
  - Liquid Removal
  - Bottle-Type and Pipe-Type Holders
  - Material Selection
  - Welding
  - Instrumentation
  - Regulation
- Compressor Station Operation and Maintenance
  - Procedures for Gas Compressor Units
  - Inspection and Testing of Relief Devices
  - Isolation of Equipment for Maintenance and Alteration
  - Inspection and Testing of Pipe-Type and Bottle-Type Holders
  - Storage of Combustible Materials
  - Inspection and Testing of Pressure Limiting and Regulating Stations
  - Prevention of Accidental Ignition

## State Air Resources Board

- Engine Exhaust Emissions

## State Fish and Game Department

- Oil Spills in Rivers, Streams, and Oceans Along the Coast
- Threats to Rare and Endangered Species of Wildlife

**State Water Quality Control Board**

• Water Discharges

**State Coastal Commission**

• Facilities Constructed in the Coastal Zone, in Addition to Regulations Imposed by Cities and Counties

# AMERICAN NATIONAL STANDARDS INSTITUTE (ANSI)

1430 Broadway
New York, NY 10018
Phone: (212) 354-3300

**ANSI B1.1 – Unified Inch Screw Threads**

**ANSI/ASME B1.20.1 – Pipe Threads, General Purpose (Inch)**

**ANSI B16.5 – Steel Pipe Flanges and Flanged Fittings, Including Ratings for Class 150, 300, 400, 600, 900, 1500, and 7500**

**ANSI B16.10 – Face-to-Face and End-to-End Dimensions of Ferrous Valves**

**ANSI B16.11 – Forged Steel Fittings, Socket Welding and Threaded**

**ANSI B16.15 – Cast Bronze Threaded Fittings, Class 125 and 250**

**ANSI B16.20 – Ring-Joint Gaskets and Grooves for Steel Pipe Flanges**

**ANSI B16.21 – Nonmetallic Flat Gaskets for Pipe Flanges**

ANSI B16.22—Wrought Copper and Bronze Solder-Joint Pressure Fittings

ANSI B16.25—Butt Welding Ends

ANSI B16.34—Steel Valves, Flanged and Butt-Welding

ANSI B18.2.1—Square and Hex Bolts and Screws

ANSI B18.2.2—Square and Hex Nuts

ANSI B18.22.1—Plain Washers

ANSI B19.1—Air Compressor (Draft Safety Standards)

ANSI B19.3—Compressors for Process Industries (Safety Standards)

ANSI B31.1—Power Piping

ANSI B31.3—Chemical Plant and Petroleum Refinery Piping

ANSI B31.8—Gas Transmission and Distribution Piping Systems

ANSI C2—National Electrical Safety Code

ANSI/ASME PTC9—Performance Test Code—Displacement Compressors, Vacuum Pumps, and Blowers

ANSI/ASME PTC10—Performance Test Code—Compressors and Exhausters

## AMERICAN PETROLEUM INSTITUTE (API)

1200 L Street, N. W.
Washington, D.C. 20005
Phone: (202) 682–8000

API 7B-11C—Internal Combustion Reciprocating Engines for Oil Field Service

API SPEC 11P—Packaged, High Speed, Separable,
Engine-Driven, Reciprocating Gas Compressors

API RP 550—Installation of Refinery Instruments and
Control Systems

API 616—Type H Industrial Combustion Gas Turbines for
Refinery Services

API 617—Centrifugal Compressors for General Refinery
Services

API 618—Reciprocating Compressors for General
Refinery Services

API 650—Welded Steel Tanks for Oil Storage

API 2000—Venting Atmospheric and Low Pressure Storage
Tanks (Nonrefrigerated and Refrigerated)


## AMERICAN SOCIETY FOR TESTING AND
## MATERIALS (ASTM)

1916 Race Street
Philadelphia, PA 19103
Phone: (215) 299-5400


ASTM A-36—Structural Steel

ASTM A-53—Pipe, Steel, Black, and Hot-Dipped Zinc
Coated, Welded and Seamless

ASTM A-106—Seamless Carbon Steel Pipe for High-Temperature
Service

ASTM A-134—Pipe Steel Electric-Fusion (Arc)—Welded (size
NPS 16 and over), Spec for

ASTM A-135—Electric-Resistance-Welded Steel Pipe

ASTM A-139—Electric-Fusion (Arc)-Welded Steel Pipe (Sizes 4-inch and over)

ASTM A-155—Electric-Fusion-Welded Steel Pipe for High-Pressure Service *Discontinued*—See A671, A672, A691

ASTM A-181—Forgings, Carbon Steel for General-Purpose Piping

ASTM A-193—Alloy-Steel and Stainless Steel Bolting Materials for High-Temperature Service

ASTM A-194—Carbon and Alloy Steel Nuts and Bolts for High-Pressure and High-Temperature Service

ASTM A-197—Cupola Malleable Iron

ASTM A-216—Carbon Steel Castings Suitable for Fusion Welding for High-Temperature Service

ASTM A-234—Piping Fittings of Wrought Carbon Steel and Alloy Steel for Moderate and Elevated Temperatures

ASTM A-307—Carbon Steel Externally Threaded Standard Fasteners

ASTM A-325—High-Strength Bolts for Structural Steel Joints

ASTM A-381—Metal Arc-Welded Steel Pipe for Use with High-Pressure Transmission Systems

ASTM A-395—Ferritic Ductile Iron Pressure-Retaining Casting for Use at Elevated Temperatures

ASTM A-490—Heat-Treated Steel Structural Bolts, 150 ksi (1035 MPa) Minimum Tensile Strength

ASTM A-539—Electric-Resistance-Welded Coiled Steel Tubing for Gas and Fuel Oil Lines

ASTM A-605—Pressure Vessel Plates, Alloy Steel, Quenched and Tempered Nickel-Cobalt-Molybdenum-Chromium

ASTM A-615—Deformed and Plain Billet—Steel Bars for Concrete Reinforcement

ASTM A-671—Electric Fusion-Welded Steel Pipe for Atmospheric and Lower Temperatures

ASTM A-672—Electric-Fusion-Welded Steel Pipe for High-Pressure Service at Moderate Temperatures

ASTM A-691—Carbon and Alloy Steel Pipe, Electric-Fusion-Welded for High-Pressure Service at High Temperatures

ASTM B-42—Seamless Copper Pipe, Standard Sizes

ASTM B-62—Composition Bronze or Ounce Metal Castings

ASTM B-75—Seamless Copper Tube

ASTM C-31—Making and Curing Concrete Test Specimens in the Field

ASTM C-33—Concrete Aggregates

ASTM C-39—Compressive Strength of Cylindrical Concrete Specimens

ASTM C-76—Reinforced Concrete Culvert, Storm Drain, and Sewer Pipe

ASTM C-94—Ready-Mix Concrete

ASTM C-150—Portland Cement

ASTM D-1751—Preformed Expansion Joint Fillers for Concrete Paving and Structural Construction (Nonextruding and Resilient Bituminous Types)

# AMERICAN SOCIETY OF
# MECHANICAL ENGINEERS (ASME)

345 East 47th Street
New York, NY 10017
Phone: (212) 705-7722

## ASME Boiler and Pressure Vessel Code, Section VIII, Division 1–Unfired Pressure Vessels

- UG 126 – Pressure Relief Valves
- UG 127 – Nonreclosing Pressure Relief Devices
- UG 131 – Certification of Capacity of Pressure Relief Valves
- UG 133 – Determination of Pressure Relieving Requirements
- UG 135 – Installation
- Appendix M-5 – Stop Valves Between Pressure Relieving Device and Vessel
- Appendix M-6 – Stop Valves on the Discharge Side of a Pressure Relieving Device

**ASME PTC7.1 – Performance Test Code – Displacement Pumps**

**ASME PTC8.2 – Performance Test Code – Centrifugal Pumps**

**ASME PTC9 – Performance Test Code – Displacement Compressors, Vacuum Pumps, and Blowers**

**ASME PTC10 – Performance Test Code – Compressors and Exhausters**

# INSTITUTE OF ELECTRICAL AND
# ELECTRONICS ENGINEERS (IEEE)

345 East 47th Street
New York, New York 10017
Phone: (212) 705-7900

**IEEE Std. 80-1976 – IEEE Guide for Safety in Substation Grounding**

# INSTRUMENT SOCIETY OF AMERICA (ISA)

67 Alexander Drive
P.O. Box 12277
Research Triangle Park, NC 27709
Phone: (919) 549-8411


**S75.01—Control Valve Sizing Equations (Replaces S39.3 Control Valve Sizing Equations for Compressible Fluids)**


# NATIONAL FIRE PROTECTION ASSOCIATION (NFPA)

Batterymarch Park
Quincy, Massachusetts 02269
Phone: (617) 770-3000


**NFPA Code 1—Fire Prevention Code**

**NFPA Code 3M—Health Care Emergency Preparedness**

**NFPA Code 9—Training Reports and Records, Recommended Practice**

**NFPA Code 10—Portable Fire Extinguishers**

**NFPA Code 10L—Model Enabling Act, Portable Fire Extinguishers**

**NFPA Code 11—Low Expansion Foam and Combined Agent Systems**

**NFPA Code 11A—Medium and High Expansion Foam Systems**

**NFPA Code 12—Carbon Dioxide Extinguishing Systems**

**NFPA Code 12A—Halon 1301 Fire Extinguishing Systems**

**NFPA Code 12B – Halon 1211 Fire Extinguishing Systems**

**NFPA Code 13 – Sprinkler Systems, Installation**

**NFPA Code 13A – Inspection, Testing and Maintenance of Sprinkler Systems**

**NFPA Code 13E – Fire Department Operations in Properties Protected by Sprinklers and Standpipe Systems**

**NFPA Code 14 – Standpipe and Hose Systems, Installation of**

**NFPA Code 15 – Water Spray Fixed Systems for Fire Protection**

**NFPA Code 16 – Deluge Foam-Water Sprinkler System**

**NFPA Code 17 – Dry Chemical Extinguishing Systems and Foam-Water Spray Systems**

**NFPA Code 18 – Wetting Agents**

**NFPA Code 19B – Respiratory Protective Equipment for Fire Fighters**

**NFPA Code 20 – Centrifugal Fire Pumps**

**NFPA Code 22 – Water Tanks for Private Fire Protection**

**NFPA Code 24 – Installation of Private Fire Service Mains and their Appurtenances**

**NFPA Code 30 – Flammable and Combustible Liquids Code**

**NFPA Code 31 – Oil Burning Equipment**

**NFPA Code 33 – Spray Application**

**NFPA Code 34 – Dipping and Coating Processes Using Flammable or Combustible Liquids**

**NFPA Code 37 – Stationary Combustion Engines and Gas Turbines**

**NFPA Code 49—Hazardous Chemicals Data**

**NFPA Code 51B—Cutting and Welding Processes**

**NFPA Code 54—National Fuel Gas Code**

**NFPA Code 58—Liquefied Petroleum Gases, Storage and Handling**

**NFPA Code 59—Liquefied Petroleum Gases at Utility Gas Plants, Storage and Handling**

**NFPA Code 59A—Liquefied Natural Gas, Storage and Handling**

**NFPA Code 68—Explosion Venting, Guide**

**NFPA Code 69—Explosion Prevention Systems**

**NFPA Code 70—National Electrical Code (NEC)**

- NEC Chapter 1—General
  - Article 100—Definitions
  - Article 110—Requirements for Electrical Installations
- NEC Chapter 2—Wiring Design and Protection
- NEC Chapter 3—Wiring Methods and Materials
- NEC Chapter 4—Equipment for General Use
- NEC Chapter 5—Special Occupancies
- NEC Chapter 6—Special Equipment
- NEC Chapter 7—Special Conditions
- NEC Chapter 8—Communications Systems
- NEC Chapter 9—Tables and Examples

**NFPA Code 70B—Electrical Equipment Maintenance**

**NFPA Code 70L—Model State Law Providing for Inspection of Electrical Installations**

**NFPA Code 71—Central Station Signaling Systems, Installation, Maintenance, and Use of**

**NFPA Code 72A—Local Protective Signaling Systems, Installation, Maintenance and Use of**

NFPA Code 72B—Auxiliary Protective Signaling Systems for Fire Alarm Service Installation, Maintenance, and Use of

NFPA Code 72C—Remote Station Protective Signaling Systems Installation, Maintenance and Use of

NFPA Code 72D—Proprietary Protective Signaling Systems, Installation, Maintenance and Use of

NFPA Code 72E—Automatic Fire Detectors

NFPA Code 77—Static Electricity

NFPA Code 78—Lightning Protection Code

NFPA Code 80—Fire Doors and Windows

NFPA Code 85A—Prevention of Furnace Explosions in Fuel Oil and Natural Gas-Fired Single Burner Boiler-Furnaces

NFPA Code 85B—Prevention of Furnace Explosions in Natural Gas-Fired Multiple Burner Boiler-Furnaces

NFPA Code 214—Water Cooling Towers

NFPA Code 220—Building Construction, Standard Types

NFPA Code 231—General Storage, Indoor

NFPA Code 321—Basic Classification of Flammable and Combustible Liquids

NFPA Code 325M—Properties of Flammable Liquids, Gases, and Volatile Solids

NFPA Code 327—Cleaning or Safeguarding of Small Tanks and Containers Using Flammable and Combustible Materials

NFPA Code 385—Tank Vehicles for Flammable and Combustible Liquids

NFPA Code 491M—Hazardous Chemical Reactions

NFPA Code 493—Intrinsically Safe Apparatus and Associated Apparatus for Use in Class I, II, and III, Division 1 Hazardous Locations

NFPA Code 495—Manufacture, Transportation, Storage and Use of Explosive Materials

NFPA Code 496—Purged and Pressurized Enclosures for Electrical Equipment in Hazardous (Classified) Locations

NFPA Code 498—Explosives, Motor Vehicle Terminals

# Appendix B

# LIST OF SYMBOLS AND ABBREVIATIONS

## Symbols

$A$ = Area of cylinder, in² (mm²)

$A_C$ = Area of crank end of piston, in² (m²) or (mm²)

$A_H$ = Area of head end of piston, in² (m²) or (mm²)

$A_R$ = Area of piston rod, in² (m²) or (mm²)

$c_p$ = Heat capacity at constant pressure, Btu/(lb•°F) [J/(kg•°C)]

$c_V$ = Heat capacity at constant volume, Btu/(lb•°F) [J/(kg•°C)]

$Cl$ = Cylinder clearance factor, percent

$d$ = Relative density, decimal

$D$ = Diameter of pipe, inches (millimetres)

$E$ = Longitudinal joint factor, decimal

$E_V$ = Volumetric efficiency factor, percent

$f_n$ = Natural frequency, Hz

$F$ = Construction type design factor, decimal

$F_C$ = Compression force in rod, lbf (N)

$F_{pV}$ = Supercompressibility factor, decimal

$F_T$ = Tension force in rod, lbf (N)

$g$ = Acceleration of gravity, 32.2 ft/s² (9.81 m/s²)

$H_a$ = Adiabatic head, ft•lbf/lb (J/kg)

$H'$ = Polytropic head, ft•lbf/lb (J/kg)

$k$ = Ratio of specific heat capacities, decimal

$k_m$ = Mean ratio of specific heat capacities, decimal

$K$ = Spring constant of soil mass, lbf/ft (kN/m)

$l$ = Length of piston stroke, inches (millimetres)

$L$ = Volumetric efficiency correction factor, percent

$L_f$ = Load factor, percent

$M$ = Molar mass of gas, lb/(lb·mol) [kg/mol]

$N$ = Rotational speed, r/min

$n$ = Polytropic exponent, decimal

$p$ = Design pressure, lbf/in² (kPa) gauge

$p_a$ = Reference base pressure, lbf/in² (kPa) absolute

$p_{amb}$ = Atmospheric pressure, lbf/in² (kPa) absolute

$p_b$ = Base pressure, lbf/in² (kPa) absolute

$p_d$ = Discharge pressure, lbf/in² (kPa) absolute

$p_f$ = Final pressure, lbf/in² (kPa) gauge

$p_i$ = Initial pressure, lbf/in² (kPa) gauge

$p_m$ = Mean effective pressure, lbf/in² (kPa) absolute

$p_s$ = Suction pressure, lbf/in² (kPa) absolute

$P$ = Power, bhp (kW)

$PD$ = Piston displacement, ft³/min (m³/s)

$P_i$ = Power rate per unit of volume, bhp/(MMcf/d)

$P_s$ = Compressor shaft power, hp (kW)

$q$ = Volume rate of flow at standard conditions, scf/h [m³/s (st)]

$q_i$ = Actual volume rate of flow, ft³/min (m³/s)

$q_m$ = Mass rate of flow, lb/h (kg/h)

$Q$ = Volume rate of flow, MMcf/d (m³/d or L/s)

$Q_H$ = Heat load, Btu/h (W)

$Q_{st}$ = Volume rate of flow at standard conditions, MMscf/d [m³/d (st) or m³/s (st)]

$R_c$ = Specific gas constant, ft·lbf/(lb·°R) [MJ/(kg·K)]

$R_f$ = Reliability factor, percent

$S$ = Yield strength of material, lbf/in² (kPa) gauge

t = Wall thickness of pipe, inches (millimetres)

$t$ = Time, s, min, h, yr

$t_1$ = Inlet temperature, °F (°C)

$t_2$ = Outlet temperature, °F (°C)

T = Temperature derating factor, decimal

$T$ = Flowing temperature, °R (K)

$T_a$ = Adiabatic temperature, °R (K)

$T_b$ = Base temperature, °R (K)

$T_d$ = Discharge temperature, °R (K)

$T_s$ = Suction temperature, °R (K)

$U_f$ = Use factor, percent

$v$ = Linear piston speed, ft/min (m/s)

$V$ = Volume, ft³ (L)

$W_f$ = Weight of machine and foundation, lbf (kN)

$W_s$ = Weight of soil system, lbf (kN)

$Z$ = Compressibility factor, decimal

$Z_d$ = Compressibility factor at discharge conditions, decimal

$Z_s$ = Compressibility factor at suction conditions, decimal

$\Delta p_r$ = Peak-to-peak pressure pulsation amplitude, percent

$\eta_a$ = Adiabatic compression efficiency, decimal

$\eta_m$ = Mechanical efficiency, decimal

$\eta_p$ = Polytropic compression efficiency, decimal

$\rho$ = Mass density, lb/scf [kg/m³ (st)]


## Abbreviations

| | |
|---|---|
| ACFM: | Actual cubic feet per minute |
| A.G.A.: | American Gas Association |
| ANSI: | American National Standards Institute |
| API: | American Petroleum Institute |
| ASME: | American Society of Mechanical Engineers |
| ASTM: | American Society for Testing and Materials |
| BMEP: | Brake mean effective pressure |
| BOCA: | Building Officials Conference of America |
| CFR: | Code of Federal Regulations |
| CRT : | Cathode-ray tube |
| CU: | Coefficient of utilization |
| DOT : | Department of Transportation |
| EPA: | Environmental Protection Agency |
| ESD: | Emergency shutdown |
| FM: | Factory Mutual |
| IEEE: | Institute of Electrical and Electronics Engineers |
| IES: | Illuminating Engineering Society |
| ISA: | Instrument Society of America |
| ISO: | International Organization for Standardization |
| LEL: | Lower explosive limit |
| LFL: | Lower flammable limit |
| LLF: | Light loss factor |
| MAOP: | Maximum allowable operating pressure |
| MEP: | Mean effective pressure |
| MTB: | Materials Transportation Bureau |
| MWP: | Maximum working pressure |
| NEC: | National Electrical Code |
| NEMA: | National Electrical Manufacturers Association |
| NESA: | National Electrical Safety Code |
| NFPA: | National Fire Protection Association |
| NPS: | Nominal pipe size |

OPSR:      Office of Pipeline Safety Regulation
OSHA:      Occupational Safety and Health Administration
PSD:      Prevention of Significant Deterioration
SCADA:      Supervisory Control and Data Acquisition
SGA:      Southern Gas Association
SPCC:      Spill Prevention Control and Countermeasure
UAERR:      Uniform Annual Equivalent Revenue Requirements
UL:      Underwriters Laboratories

# GLOSSARY

An attempt has been made to explain terms specific to compressor applications where they appear in the text. However an expanded definition of certain compressor-related terms is offered in this glossary.

**Air/fuel ratio** is the mass ratio of air to fuel taken into an engine for combustion.

**Ampacity** is the electrical current in amperes a conductor can carry continuously under the conditions of use without exceeding its temperature rating.

**Availability factor** is the ratio of the time that a machine is capable of operating and is available for operation – whether or not it is actually operated – to the total hours in an operating period (8 760 h = 1 year).

**Back-pressure regulator** is an automatic valve arrangement whereby the fluid pressure is maintained at a constant value on the upstream side of the device.

**Brake mean effective pressure (bmep)** is the mep which, if acting on the piston of an engine, would develop power equivalent to the brake horsepower. The bmep provides an accurate means for comparison of engine performance and is preferred to the indicated mep. The bmep is calculated from the measured bhp data.

**Bus (or bus-bar)** is a flat copper or aluminum strip used as a conductor in electrical construction. The sizing is based upon current density of about 800 to 1 200 A/in² (1.25 to 1.85 A/mm²) for copper, and somewhat less for aluminum.

**Combined cycle** is a power cycle combining two or more different thermodynamic cycles; for example, hot gas-turbine exhaust may be used to generate steam to drive a steam turbine.

**Compression Cycles:**

**Adiabatic (isentropic) compression** takes place when there is not heat added to or removed from the system. Compression follows the formula $p_1V_1{}^k = p_2V_2{}^k$, where exponent $k$ is the ratio of the specific heat capacities. Although an adiabatic cycle is never totally obtained in practice, it is approached typically with most positive-displacement machines and is generally the base to which they are referred.

**Isothermal compression** takes place when the temperature is kept constant as the pressure increases, requiring continuous removal of heat generated during compression. Compression follows the formula $p_1V_1 = p_2V_2$. However, in practice it is never possible to remove the heat of compression as rapidly as it is generated.

**Polytropic compression** is a compromise between the two basic processes, the adiabatic and the isothermal. It is primarily applicable to dynamic continuous-flow machines such as centrifugal or axial compressors. Compression follows the formula $p_1V_1{}^n = p_2V_2{}^n$, where exponent $n$ is experimentally determined for a particular type of machine. It may be lower or higher than the exponent $k$ used in adiabatic cycle calculations.

**Compression efficiency** is the ratio of the theoretical work requirement (using a stated process) to the actual work required to compress a given quantity of gas. It accounts for the gas friction losses, internal leakage and other variations from the idealized thermodynamic process.

**Cylinder clearance** is the volume remaining in the end of a cylinder that is not swept by the movement of the piston. It includes the space between piston and head at the end of the compression stroke, the space under valves, etc., and is expressed as a percentage of the piston displacement per stroke. Clearance can be different for the two ends of a double-acting cylinder, in which case an average value is used.

**Decibel (dB)** is one-tenth of a bel (B), the number of decibels denoting the ratio of the two amounts of power being ten times the logarithm to the base ten of this ratio. When voltages, currents, sound pressure or analogous quantities are used in place of power, the ratio of the two amounts becomes 20 times the log to the base ten.

**Decibels referred to the A-scale (dBA)** is the sound pressure level equivalent to that obtained using an instrument equipped with an A-scale weighting network. The A-scale falls off sharply in the low frequencies, corresponding to response characteristics of the human ear, thus providing a useful standard for annoyance or potential hearing loss. Other weighting scales employed in engineering studies are B and C.

**Engine indicator** is a device for plotting cylinder pressure as a function of piston stroke. The resulting $pV$ analog provides both a measure of the work done in a reciprocating engine, a pump or compressor, and a means for analyzing their performance.

**Failsafe** is a design feature incorporated in control systems to minimize consequential damages after a control component failure or malfunction.

**Fan laws** define the performance of dynamic machines at points other than the rated flow and speed point. Since the $Q/N$ value is constant, the volume flow rate is directly proportional to rotational speed ($Q_2N_1 = Q_1N_2$) and the head developed by an impeller at a constant flow rate is proportional to the square of the speed ($H_2N_1^2 = H_1N_2^2$). Since the power requirement is directly proportional to the product of the flow rate and the head, the gas power is proportional to the cube of the speed ($P_2N_1^3 = P_1N_2^3$), assuming that both head and the flow rate are permitted to rise.

**Fuller's earth** is the trade name for a siliceous adsorbent used as a filter media for lubricating oils, etc.

**Gas generator** is the section of a gas-turbine that provides the high temperature gases needed to drive the power turbine.

**Gas turbine** is an engine powered by the reaction and/or impulse of high temperature gases against the curved blading affixed to the rotor.

**Gas-turbine cycle (Brayton, or Joule)** is a power cycle consisting of adiabatic compression of air (or other working fluid), heating at constant pressure, and adiabatic expansion. Because the expanding gases are hotter, the work obtainable in the expansion process is greater than the input into compression.

**Grout** is a thin mortar, fluid enough to be used for filling small spaces to form concrete.

**Halon** is a family of halogenated hydrocarbon gases used in fire extinguishing systems covered by NFPA Standards 12A and 12B. Halon 1301 ($CF_3Br$), because of its low toxicity, is used in many total flooding systems protecting electronic computer/data processing equipment in the U.S.A.; Halon 1211 ($CF_2BrCl$) is used for similar applications in

Europe, but only with precautions, such as time delays, not recognized in NFPA Standards 12A and 12B.

**Head** is a common name for the change in the specific energy. When applied to compression of gas, it represents the amount of energy per unit mass that must be added to gas to produce the desired pressure and/or velocity increase. (In inch-pound system of units based on mass density (lb/ft³), the unit for head most frequently used is ft·lbf/lb; when units are based on specific weight or weight density (lbf/ft³), the unit for head simplifies to ft.) The concept is very useful, since the gas horespower requirement is directly proportional to the mass flow rate multiplied by the head and inversely proportional to the efficiency.

**Heat rate** is the fuel consumption by an engine expressed as input energy per unit of work (at the output shaft).

**Heating values (of a gaseous fuel):**

While gas energy sales normally involve use of the Gross or Total Heating Value ($H_s$) energy-related calculations in prime mover technology are based on the Net Heating Value ($H_i$). The ratio of net to gross for pipeline quality gas is about 90.1%; therefore, efficiency calculated using $H_i$ will be about 10% higher than that using $H_s$.

**Gross heating value** is the quantity of heat produced by the combustion of a unit of gas in air under constant pressure – the air being at the same temperature and pressure as the gas – after cooling of the combustion products to the initial temperature of the gas and after condensation of the water vapor created by the combustion to the liquid state.

**Net heating value** is less than the gross heating value by an amount equal to the heat of vaporization of water formed in the combustion of gas at the initial temperature of gas and air.

**Horsepower (hp)** is a unit of power equal to 33 000 foot-pounds force per minute (1hp=746 W, approx; not to be confused with metric horsepower, 1 PS=735 W, approx.).

**Theoretical horsepower** is the power theoretically required to compress and deliver a given quantity of gas in accordance with a specified process.

**Gas (or Indicated) horsepower (ihp)** is the actual power requirement for compression and delivery of gas at particular conditions, including all thermodynamic, leakage and gas friction losses, but excluding mechanical friction.

**Brake horespower (bhp)** is the total power requirement of a compressor, consisting of gas horsepower and all mechanical friction and power transmission losses; it equals the net power delivered by an engine.

**Friction horsepower** is the difference between the indicated and the brake horsepower, and includes the negative loop (pumping loss) of the indicator diagram.

**Horsepower hour (hph)** is amount of work or energy equal to $1.98 \times 10^6$ ft•lbf (2.685 MJ approx.); this amount is produced in 1 hour by an engine developing one horsepower (1 hp = 746 W).

**Internal combustion engine** may be either of a reciprocating piston type or a rotary, where the theoretical combustion process may take place at a constant volume (Otto cycle), at constant pressure (Diesel cycle), or at combination of the two ("dual-fuel"). The principal differences are in the timing of the fuel introduction during the cycle and the method of ignition.

**Intrinsically safe loop** is comprised of equipment and wiring that is incapable of releasing sufficient electrical energy under normal or abnormal conditions to cause ignition of a specific hazardous atmospheric mixture; therefore, provisions of Articles 500 to 517 of the National Electrical Code need not apply to such an installation.

**Lower explosive limit (LEL)** is the limit for concentration of flammable gas or vapor in air, expressed as percent by volume, below which the mixture is too lean to support propagation of a combustion wave away from the source of ignition no matter how much energy is put into the mixture.

**Load factor ($L_f$)** is the ratio, generally in percent, of power developed by an engine (or a station) to the rated power.

**Manway** is an opening into a tank, boiler, furnace or other equipment through which a person may enter.

**Mean effective pressure (mep)** is the pressure on a piston which when multiplied by the area of the piston and the stroke gives the net work done on the piston in one cycle of events.

**Nominal pipe size (NPS)** designates a method of identifying the size of steel pipe without compromising the actual diameter data. Thus, nominal 1 in pipe with the actual external diameter of 1.315 in (33.40 mm) becomes NPS 1; nominal 14 in pipe with internal diameter of 13.250 in (336.55 mm) becomes NPS 14.

**Overpressure protection** is a system installed to prevent pressure in a piping system from exceeding a predetermined limit. The system may incorporate back-pressure and monitor regulators, safety and relief valves, rupture discs, etc.

**Reciprocating engine** is a heat engine, generally of internal combustion type, in which the back-and-forth motion of the piston is transformed into a rotating motion of the crankshaft.

**Regenerative cycle** is a gas turbine cycle wherein the output of the inlet air compressor is preheated by the turbine exhaust before it is fed into the combustor. A regenerative cycle has a higher thermal efficiency than does a simple gas turbine cycle.

**Relief valve** is a safety valve that opens as needed to vent the fluid out of a system (pipe and vessels) to prevent overpressure.

**Rotary engine (Wankel)** is an internal combustion engine incorporating a three-sided rotor revolving in timed relationship within a two-lobed rotor housing.

**Scavenging air** forces products of combustion out of a 2-stroke engine cylinder, and charges the cylinder with fresh air before compression and ignition occur.

**Stoichiometric combustion** is a reaction in which all fuel is burned with the theoretically correct amount of air. Neither unburned fuel nor free oxygen are present in the stack gases; carbon dioxide ($CO_2$) is at a maximum value in the combustion products.

**Steam turbine** is an engine powered by the reaction and/or impulse of high temperature steam against the curved blading affixed to the rotor.

**Supercharging** is a method used to increase the pressure, and thereby the amount of charge per cycle, above that of a normally aspirated internal combustion engine; it permits more fuel to be burned and is a practical means to greater engine power.

**Thermodynamic (Absolute) temperature** is the temperature referenced to absolute zero, a theoretical point at which all molecular motion stops. On the Fahrenheit scale this corresponds to minus 459.67 °F; on the Celsius scale to minus 273.15 °C. When numerical values have to be positive, either Rankine or Kelvin scales are used with the absolute zero corresponding to 0°R or 0 K.

**Thermal efficiency** is the ratio of the work produced to the energy supplied. Gas engine efficiencies normally are based on the net heating value of the fuel.

**Torque** is the turning moment, or force exerted at a radius, and is frequently treated as a measure of the output of an engine (1 lbf•ft=1.356 N•m, approx.).

**Turbocharging** is a method of supercharging an engine using an exhaust gas-driven air compressor.

**Upper explosive limit (UEL)** is the limit for concentration of flammable gas or vapor in air, expressed as percent by volume, above which the mixture is too rich to support propagation of a combustion wave away from the source of ignition no matter how much energy is put into the mixture.

**Use factor ($U_f$)** is the ratio, generally expressed in percent, of the total hours operated to the total calendar hours in an operating period (usually a month).

**Volume/speed characteristic ($Q/N$ value)** is the ratio of the optimum stable flow rate per minute (at inlet pressure and temperature) to the operating speed (in r/min) of the impeller. The value represents the volume capacity of an impeller (actual cubic feet, or actual cubic metres, per revolution).

**Volumetric efficiency** is the ratio of the volume (measured at the inlet conditions) actually delivered to the piston displacement of a reciprocating compressor.

# BIBLIOGRAPHY

American Gas Association, *Engine Analyzer Signature Directory,* Arlington, VA: American Gas Association, 1978. [Catalog No. XF0478]

American Gas Association Operating Section Compressor Committee, *Classification of Gas Utility Areas for Electrical Installations,* 4th edition, Arlington, VA: American Gas Association, 1973. [Catalog no. XF0277]

American Gas Association, *Gas Engineers Handbook,* New York: The Industrial Press, 1974.

American Gas Association, *Gas Measurement Manual,* Arlington, VA: American Gas Association, 1981. [Catalog no. XQ1081]

American Gas Association, *Orifice Metering of Natural Gas,* Arlington, VA: American Gas Association, 1978. [Catalog no. XQ0178]

American Society of Mechanical Engineers, *ASME Guide for Gas Transmission and Distribution Piping Systems,* 5th edition, New York: American Society of Mechanical Engineers, 1983.

American Telephone and Telegraph Company, *Engineering Economy: A Manager's Guide to Economic Decision Making,* 3rd edition, New York: McGraw-Hill Book Company, 1977.

Anderson, J. B., "Documentation of the Benedict-Webb-Rubin Mark 2 Computer Program," American Gas Association, *1972 Operating Section Proceedings,* pp. T75–T83, Discussion p. T84. [Catalog no. X59972]

Arya, Suresh C., Drewyer, Roland P., and Pincus, George, "Foundation Design for Reciprocating Compressors," *Hydrocarbon Processing,* vol. 56, no. 5, May 1977, pp 223–232.

Axford, Mark and Stinger, Dan, "Combined Cycle Operation for Pipeline Compressor Stations," *Turbomachinery International,* vol. 21, no. 6, July/August 1980, pp. 17–22.

Barkan, D. D., *Dynamics of Bases and Foundations*, New York: McGraw-Hill Book Company, 1962.

Beeman, Donald, Editor, *Industrial Power Systems Handbook*, New York: McGraw-Hill Book Company, 1955.

Bolt Beranek and Newman, Inc., *Noise Control for Reciprocating and Turbine Engines Driven by Natural Gas and Liquid Fuel*, New York [now Arlington, VA]; American Gas Association, December 1969.

Bowles, Joseph E., *Foundation Analysis and Design*, New York: McGraw-Hill Book Company, 1968.

Chlumsky, Vladimir, *Reciprocating and Rotary Compressors*, London: E & FN Spon Ltd., 1965.

Dimoplon, William, "What Process Engineers Need to Know about Compressors," *Hydrocarbon Processing*, vol. 57, no. 5, May 1978, pp 221–227.

Dranchuk, P. M. and Abou-Kassem, J. H., "Calculation of Z Factors for Natural Gases Using Equations of State," *The Journal of Canadian Petroleum Technology*, vol. 14, no. 3, July-September 1975, pp 34–36.

Edmister, Wayne C., *Applied Hydrocarbon Thermodynamics*, Houston: Gulf Publishing Company, 1961, 2 vols.

Gartmann, Hans, Editor, *DeLaval Engineering Handbook*, 3rd edition, New York: McGraw-Hill Book Company, 1970.

Gas Processors Suppliers Association and Gas Processors Association, *Engineering Data Book*, 9th Edition, Tulsa: Gas Processors Suppliers Association, 1972.

Grant, Eugene L. and others, *Principles of Engineering Economy*, 7th edition, New York: John Wiley & Sons, 1982.

Henry, Martin, F., Editor, *Flammable and Combustible Liquids Code Handbook*, Quincy, MA: National Fire Protection Association, 1981.

Kannappan, S., "Determining Centrifugal Compressor Piping Loads," *Hydrocarbon Processing*, vol. 61, no. 2, February 1982, pp. 91–93.

Katz, D. L., *Handbook of Natural Gas Engineering*, New York: McGraw-Hill Book Company, 1959.

Kaufman, John E., Editor, *IES Lighting Handbook*, New York: Illuminating Engineering Society of North America, 1981. Vol. 1, Reference Volume. Vol. 2, Application Volume.

Kennedy, J. L., "What You Should Know About Reciprocating Compressors—Part 1," *Oil and Gas Journal*, vol. 65, no. 46, November 13, 1967, pp 105–107; Part 2, no. 48, November 27, pp 95–97; Part 3 no. 49, December 4, pp 72–74; Part 4, no. 51, December 18, pp 76–78; Part 5, vol. 66, no. 2, January 8, 1968, pp 65–68; Part 6, no. 4, January 22, pp 76–78.

Leonards, G. A., Editor, *Foundation Engineering,* New York: McGraw-Hill Book Company, 1962.

Loomis, W. A., Editor, *Compressed Air and Gas Data,* Revised 3rd edition, Woodcliff Lake, NJ: Ingersoll-Rand Co., 1982.

Marquis, D. E., "Gas Engines Offer Potential for Energy Savings," *Oil and Gas Journal,* vol. 78, no. 52, December 29, 1980, pp 191, 194–197.

Mayer, Raymond B., *Capital Expenditure Analysis for Managers and Engineers,* Prospect Heights, IL: Waveland Press, Inc., 1978.

McPartland, J. F., Editor, *McGraw-Hill's National Electrical Code Handbook,* 18th edition, New York: McGraw-Hill Book Company, 1984.

Nesterenko, Dimitri A. and Pugh, Robert G., "Dynamic Design for Large Compressor Foundations," *Compressed Air Magazine,* vol. 74, no. 7, July 1969, pp 9–13.

Pauw, A., "A Dynamic Analogy for Foundation-Soil Systems," *Symposium Dynamic of Testing Soils,* ASTM Special Technical Publication no. 156, Philadelphia: American Society for Testing and Materials, 1953, pp 90–112.

Popan, Victor A., "How to Calculate Pressure Drop in Compressor Station Piping," *Pipeline & Gas Journal,* vol. 204, no. 6, May 1977, pp 29–33.

Rausch, Ernst, *Maschinenfundamente und andere dynamische Bauaufgaben,* 2nd edition, Berlin: Vertrieb VDI-Verlag G.m.b.H., 1943, 3 vols.

Richart, F. E., Jr., "Foundation Vibrations," *Transactions of the American Society of Civil Engineers,* vol. 127 Part 1, 1962, pp 863–925.

Rollins, John P., Editor, *Compressed Air and Gas Handbook,* 4th edition, New York: Compressed Air and Gas Institute, 1973.

Scheel, Lyman F., *Gas and Air Compression Machinery,* New York: McGraw-Hill Book Company, 1961.

Scheel, Lyman F., *Gas Machinery,* Houston: Gulf Publishing Company, 1972.

Shaw, Howard C., "Reciprocating Versus Centrifugal Compressors," *Pipe Line Industry,* vol. 54, no. 5, May 1981, pp 39–43.

Smith, Gerald W., *Engineering Economy: Analysis of Capital Expenditures,* 3rd edition, Ames, IA: Iowa State University Press, 1968.

Sowers, G. B. and Sowers, G. F., *Introductory Soil Mechanics and Foundations,* 3rd edition, New York: The Macmillan Company, 1970.

Sowers, George F., *Introductory Soil Mechanics and Foundations: Geotechnical Engineering,* 4th edition, New York: Macmillan

Publishing Co., Inc., London: Collier Macmillan Publishers, 1979.

Stermole, Franklin J., Editor, *Economic Evaluation and Investment Decision Methods,* 5th edition, Golden, CO: Investment Evaluations Corporation, 1982.

Taylor, Charles Fayette, *The Internal-Combustion Engine in Theory and Practice,* Cambridge, MA: The Technology Press of The Massachusetts Institute of Technology *and* New York: John Wiley & Sons, Inc., 1960.

Thuesen, H. G., Fabrycky, W. J., and Thuesen, G. J., *Engineering Economy*, 6th edition, Englewood Cliffs, NJ: Prentice-Hall, 1984.

Urik, Jim and Odom, Fred, "Calculator Analyzes Compressor Performance," *Oil and Gas Journal,* vol. 78, no. 2, January 14, 1980, pp 60–65.

White, John A., Agee, Marvin, H., and Case, Kenneth E., *Principles of Engineering Economic Analysis,* New York: John Wiley & Sons, Inc., 1984.

Wilkins, John, "How to Get a Permit for New Compressor Horsepower," *Pipeline & Gas Journal,* vol. 207, no. 14, December 1980, pp 56, 58, 60–62.

# INDEX

## V



## W